Agricultural Biotechnology

Agricultural Biotechnology

Hemant Rawat

Oxford Book Company
Jaipur ▪ New Delhi

Published By

Oxford Book Company

Regd. Off. "Bony Residency", Gate No. 2
Opp. Tilak Public School, Vishweswariya Nagar,
Gopalpura Road, Jaipur - 302018, India
Ph: 0141 - 2761280, Telefax: 2761381
email - oxfordbookjpr@gmail.com
website - oxfordbookcompany.com

Delhi Office: 102, 1st Floor, Satyam House,
4327/3, Ansari Road, Darya Ganj,
New Delhi - 110002, India
Ph: 011-45652440

Distributed By

KEYA INTERNATIONAL PRESS LLP

E-mail: keyainternationalpressllp@gmail.com

Published in India, 2024

ISBN : 9788189473235

Price: ₹ 1850.00

Typeset by Shivangi Computers, Jaipur

Printed and bound in India by Thomson Press Ltd.,
New Delhi

Preface

Agricultural Biotechnology is one of the most important key technologies of the 21 century. It provides great potential in the area of plant breeding. In this way, plants can be developed that provide ingredients for healthier nutrition, grow in unfavourable conditions or form substances that are otherwise produced by means of a complex chemical process. Through biotechnology not only the variety is improved, but yield also improves through growing crops having resistance to diseases, herbicides and pesticides.

Through understanding basic concepts of agricultural biotechnology and adopting improved genetically mutated seed varieties, the grower as well as the consumer is benefited a great lot. Agricultural Biotechnology" presents all the latest techniques of plant tissue culture, transformation and bioengineering at the outset and discusses in detail various issues of plant breeding, cloning, disease resistance, and herbicide and pest resistance.

The book makes elaborate presentation on radio-immunoassays, enzyme immuno-absorbent assays, genetic recombination, pharmaceutical products and solar energy that have immensely impacted the field of biotechnology.

Hemant Rawat

Contents

1

Introduction

BIOTECHNOLOGY AND PLANTS

Biotechnology is being used as a tool to give plants new traits that benefit agricultural production, the environment, and human nutrition and health. The purpose of this publication is to provide basic information about plant biotechnology and to give examples of its uses. The goal of plant breeding is to combine desirable traits from different varieties of plants to produce plants of superior quality. This approach to improving crop production has been very successful over the years.

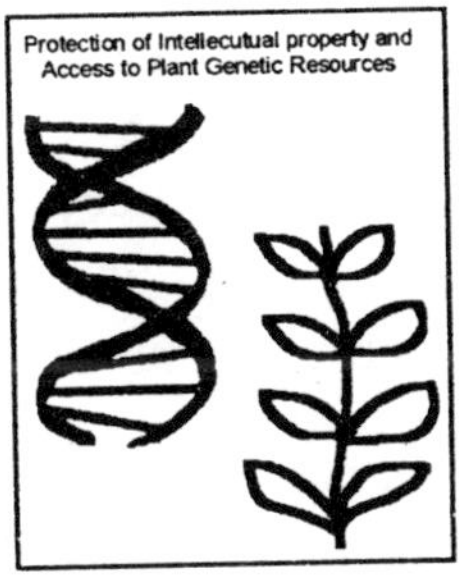

Fig. Plant breeding

For example, it would be beneficial to cross a tomato plant that bears sweeter fruit with one that exhibits increased disease resistance. To do this, it takes many years of crossing and backcrossing generations of plants to obtain the desired trait. Along the way, undesirable traits may be manifested in the plants because there is no way to select for one trait without affecting others. Another limitation of traditional plant selection is that breeding is restricted to plants that can sexually mate.

Advances in scientific discovery and labouratory techniques during the last half of the twentieth century led to the ability to manipulate the deoxyribonucleic acid (DNA) of organisms, which accelerated the process of plant improvement through the use of biotechnology.

MODERN PLANT BIOTECHNOLOGY

Plants are made of millions of cells all working together. Every cell of a plant has a complete "instruction manual" or genome (pronounced "JEE-nom") that is inherited from the parents of the plant as a combination of their genomes.

Genes are found within the genome and serve as the "words" of the instruction manual. When a cell reads a word, or in scientific terms "expresses a gene," a specific protein is produced. Proteins give an individual cell, and therefore the plant, its form and function. Genes (words) are written using the four-letter alphabet A, C, G, T. The letters are abbreviations for four chemicals called bases, which together make up DNA. DNA is universal in nature, meaning that the four chemical bases of DNA are the same in all living organisms. Consequently, a gene from one organism can function in any other organism.

The ability to move genes into plants from other organisms, thereby producing new proteins in the plant, has resulted in significant achievements in plant biotechnology that were not possible using traditional breeding practices.

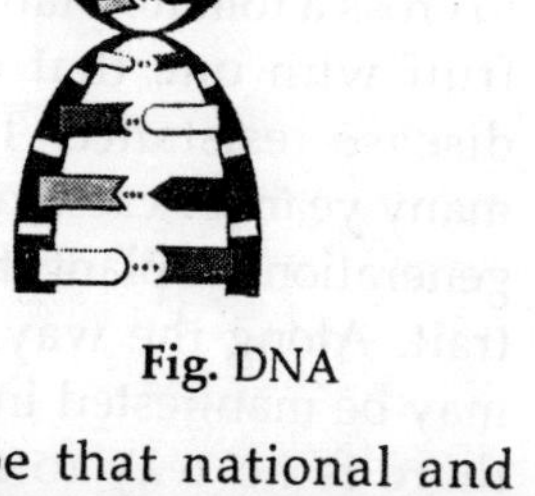

Fig. DNA

Clearly, the ultimate solution to each of these problems is reducing population growth, a difficult challenge that is further complicated by social, political, economic and religious considerations. In the hope that national and international efforts will help to stabilize world population in the next few decades, our challenge is to use the power of plant biotechnology toward the solution of the numerous problems caused by population growth by increasing productivity, by reducing crop losses, and by protecting and conserving the environment.

Plant biotechnology is not a magic bullet that will solve all of these problems, yet it is becoming abundantly clear that it is the best tool that we have and it can, if used wisely and in a timely fashion, make significant contributions. It is rather ironic that at a time when international agriculture is under increasing pressure to meet the food needs of the ever increasing population, and when plant biotechnology is beginning to make significant contributions to food productivity and environmental safety, it has become the target of well coordinated and sustained attacks by many environmental and self-appointed watchdog groups, particularly in Western Europe and in some of the developing countries. In India, we had faced similar attacks during the 1970's and 1980's.

However, after extensive public debate, protest demonstrations, court challenges and congressional hearings, a federal regulatory framework was developed that has served the public and the private interest well. It has allowed the plant biotechnology industry to grow and introduce its products into the market place. The Indian consumers and farmers have accepted and benefited from transgenic products. Transgenic crops are being grown this year on nearly 100 million acres of Indian farmland, accounting for 74% of our soybean, 71% of our cotton and 32% of our corn acreages.

It is fortunate and encouraging that China and India, the two most populous countries in the world, with increasing demand for food and worsening environmental problems, have recognized the importance of plant biotechnology in agriculture and have established active and successful research and development programmes in plant biotechnology, targeting many regional vegetable and fruit crops, in addition to such staples as wheat, rice, maize, soybean, various pulses, canola, and cotton. It is not too far-fetched to expect that within the next few years these two countries will plant the largest acreages of transgenic crops in the world.

Argentina and South Africa are two other developing countries that are increasing their planting of transgenic crops.

Indeed, at the present time, nearly 10% of the global acreage of transgenic crops is planted in the developing countries. Thus the argument that plant biotechnology is a tool of the industrialized countries for the exploitation of the developing world is no longer sustainable. It is our hope that the success of plant biotechnology in these countries will encourage similar efforts in other parts of the developing world.

It is true that the first generation of transgenic crops, which contain genes for resistance to herbicides and insects, did not provide any direct benefits to the consumer. Nevertheless, there are numerous indirect benefits, such as the reduced use of pesticides and herbicides, reduced tillage leading to soil conservation, reduced use of natural resources such as petrochemical products and water for the manufacture, transport and application of agro-chemicals, and reduced labour costs. By producing more food on the same amount of land transgenic crops promote conservation and biodiversity by saving wildlife habitats and precious forests from being converted into farmland.

Reduced use of pesticides has already shown a marked decrease in illness and death caused by pesticide poisonings in China and South Africa (nearly 500 cotton farmers in China die each year of acute pesticide poisoning). The vital role of agriculture and food production in human health and nutrition, in poverty alleviation, and in social and political stability, is well known. This was recognized as far back as 1970, when Norman Borlaug was awarded the Nobel Peace Prize for his work that led to the Green Revolution and that helped to save hundreds of millions of lives in the developing countries.

Plant biotechnology too can contribute to international peace and security by increasing food production, producing safer and healthier foods, protecting our rather finite natural resources and the environment, and improving human health. It is, therefore, morally and socially irresponsible and indefensible to prevent or delay the applications of plant biotechnology to problems of hunger, health and protection of the environment.

The opponents of plant biotechnology would have us believe that it is an unnatural and unsafe process that produces harmful products, and that it is totally different from plant breeding and selection that account for almost all of our modern crops. The indisputable fact, however, is that humans have engineered crops for nearly 10,000 years. Almost all of our major crops - such as maize, wheat, potato, tomato and others - are man made. Indeed, none of our modern crops are capable of surviving in the wild without human care.

The molecular and genetic principles of plant biotechnology and plant breeding and selection are the same. Plant biotechnology is no different from breeding and selection or for that matter from radiation and chemically induced mutation breeding, except that it is extraordinarily precise and predictable, and is not restricted by taxonomic boundaries. No compelling evidence has ever been presented to show that transgenic crops are innately different the non-transgenic products of breeding and selection.

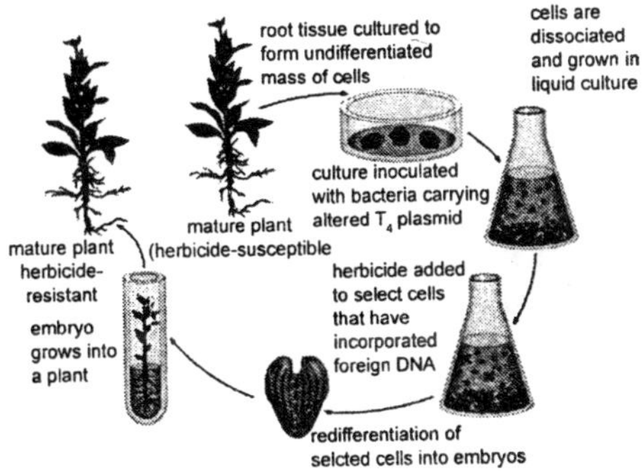

Fig. Transgenic Crops

In retrospect, however, we must share some of the blame for the perception that plant biotechnology is different from plant breeding and selection. During the 1970's and 1980's, when there was a great deal of euphoria over the production of somatic hybrids, doubled-haploid breeding lines and transgenic plants, the plant biotechnology community made a serious error in strategy and judgment when it distanced itself from breeding and selection and established a separate identity for itself.

It was us who placed the spotlight on the process and not the product. This has come to haunt us now as it has attracted undue and undeserved attention and opposition. It was also an error not to engage early in the debate on transgenic plants, and to permit the opponents of plant biotechnology to dictate the agenda.

As responsible members of the world community, and as scientists, we cannot, and should not, be silent observers of this debate. We must play an active role in the debate on plant biotechnology and make an informed and professional contribution to the public dialogue, emphasizing the many benefits of transgenic crops to human health and the environment. The opposition to transgenic foods in Europe and elsewhere is based exclusively on political and ideological differences rather than on any credible scientific evidence. On two rare occasions an attempt was made to present scientific arguments against the use of transgenic crops. These involved the allegedly harmful effect of pollen from Bt maize plants on the larvae of the Monarch butterfly, and the alleged transgene contamination of maize in Mexico.

More detailed investigations by several research groups have since refuted these claims and showed them to be of dubious scientific value. Indeed, in the Mexican maize story, the journal Nature, in an unprecedented action in its more than 100 year history, was forced to disown the paper published in its own pages.

The consumer, the farmer and the biotechnology industry have been ill served by the sustained campaign of misinformation and unsubstantiated claims of dangers to public health and the environment by transgenic crops and their products. After more than ten years and thousands of field trials in many countries, after nearly a decade of commercial plantings on hundreds of millions of acres, and after transgenic food products having being used by hundreds of millions of humans and farm animals, there is not a single documented instance of illness reported in any human or animal, or of ecological or environmental damage. What then is the basis and rationale for the many restrictions still placed on the field planting and human use of transgenic foods?

The enviable and unblemished record of transgenic crops and their products is the strongest evidence for their safety and wholesomeness. The opponents of plant biotechnology should compare this record with that of the many drugs approved for human use in the United States. In an exhaustive study published recently in the Journal of the Indian Medical Association, it was reported that 20% of the 548 drugs approved for human use during the past 25 years were later found to have serious or life-threatening side effects. Seven of the drugs possibly contributed to 1002 deaths, and 16 were forced to be withdrawn from the market. In comparison, not a single transgenic food product has ever been shown to have any harmful effects, and none has been withdrawn because of adverse reactions in humans or animals.

The plant biotechnology community has already done more for the environment and the developing countries than the self-proclaimed environmental groups and the so-called friends of the poor. Indeed, the opponents of plant biotechnology have done much harm to their professed cause by slowing down and/or preventing the planting and utilization of transgenic crops around the world. The contributions of the plant biotechnology community, on the other hand, are socially and morally responsible and of considerable humanitarian value. We have every reason to be proud of these contributions.

The rules and regulations adopted for transgenic crops in the 1980's were both prudent and necessary. At that time there were many unknowns about transgenic crops and about their possible effect on humans and the environment. There was a need to establish a database to satisfy the concerns of the general public as well as the scientific community. Three federal agencies, the United States Department of Agriculture, the Environmental Protection Agency and the Food and Drug Administration, were given oversight responsibilities for transgenic crops.

The resulting open and transparent system established in the United States has worked well and has served its purpose. It has done much to gain the confidence and support of the American public for plant biotechnology and its products. In light of the demonstrated safety of transgenic crops to humans, animals and the environment, the question must be asked whether it is any more necessary, or even advisable, to continue the expensive, time consuming and burdensome requirements for the public release of transgenic crops and their products (field trials of transgenic crops are 10-20 times more expensive than of similar plants developed by conventional means).

I propose that based on our considerable experience and on the vast amount of information gathered about the safety of transgenic crops over the past decade, it is time for our regulatory agencies to consider whether some or all of the current regulatory requirements can be gradually relaxed and ultimately suspended, except in those rare instances where there is the clear likelihood of risk to human health and the environment. Genuine concerns about gene flow and development of resistance to antibiotics, and pests or pathogens, can be met adequately with currently available and emerging technologies.

The process of deregulation of transgenic crops, controversial and difficult as it may be, needs to begin now because the continuation of the present rules and regulations is entirely unnecessary, unjustified and counterproductive. In order to be effective and acceptable, the process should be open to all points of view. The decisions, however, must be based on science and facts and not on political or ideological considerations. Nearly two decades ago, the United States played a leading and useful role in establishing the rules and regulations for the field planting, evaluation and human use of transgenic crops. It should now play a similar role in having these restrictions relaxed and removed.

It is clear that the challenges we face in the 21st century are greater than those we faced in the last century. Of all the available technologies, plant biotechnology offers the best hope for producing more and better food, fiber and pharmaceuticals, and for protecting, preserving and improving the environment for the benefit of humankind. My own confidence in plant biotechnology comes from knowing that the science behind it is sound, that it is well tested and proven, that it benefits the consumer, the farmer and the industry, and that it protects and conserves the environment.

It is for these reasons that I am convinced that plant biotechnology will within the next two decades become an integral part of the international agricultural system. With the United States, China, and lately India, three of the most populous countries in the world serving as examples, we have taken the first steps toward achieving that objective.

Introducing Genes into Plants

To genetically modify a plant, the thousands of bases of DNA comprising an individual gene are transferred into an individual plant cell where the new gene becomes a permanent part of the cell's genome. This process makes the resulting plant "transgenic." Transfer of DNA into plant cells is done using various "transformation" techniques that are the result of discoveries in basic science.

One method to transfer DNA into plants takes advantage of a system found in nature. The bacterium that causes "crown gall tumors" injects its DNA into a plant genome, forcing the plant to create a suitable environment for the bacterium to live. After discovering this process, scientists were able to "disarm" the bacterium, put new genes into it, and use the bacterium to harmlessly insert the desired genes into the plant genome.

In the "biolistic" or "gene gun" method, microscopic gold beads are coated with the gene of interest and shot into the plant cell with a burst of helium. Once inside the cell, the gene comes off the bead and integrates into the cell's genome.

It was also discovered that plant cells could be "electroporated" or mixed with a gene and "shocked" with a pulse of electricity, causing holes to form in the cell through which the DNA could flow. The cell is subsequently able to repair the holes and the gene becomes a part of the plant genome.

When using these methods, new genes are successfully introduced into only a small percentage of the cells, so scientists must be able to "pick out" or "select" the transformed cells before proceeding. This is often done by concurrently introducing an additional gene into the cell that will make it resistant to an antibiotic.

A cell that survives antibiotic treatment will most likely have received the gene of interest as well; that cell is subsequently used to propagate the new plant. There is a concern that the gene giving antibiotic resistance could naturally be transferred to bacteria once the transgenic plant is in the wild, making bacteria resistant to antibiotics that are used to fight human infection. Scientists are currently devising ways to select for transformed cells that will alleviate this issue.

Timeline of Plant Biotechnology

1700s — Naturalists identify hybrid plants

1860s — Austrian botanist and monk Gregor Mendel studies pea plants and recognizes that specific traits are passed from parents to offspring - these traits are eventually discovered to be genes

1900 — European botanists begin to improve plant productivity using genetic theories based on Mendel's work

1922 — Farmers purchase hybrid seed corn created by crossbreeding two corn varieties

1953 — Structure of DNA is discovered - marking the beginning of modern genetic research

1970s — Hybrid seeds are introduced to developing countries to increase food supplies

1973 — Genetic engineering is used to precisely manipulate bacterial DNA

1983 — First GM plant is created; a tobacco plant resistant to an antibiotic

1985 — GM plants resistant to viruses, bacteria, and insects are f'eld tested

1986 · EPA approves the release of th3e first GM crop (herbicide resistant tobacco)

1990 — First successful field trial of GM cotton (herbicide resistant)

1992 — FDA decides GM foods will be regulated as conventional foods

1994 — FlavrSavr Tomato becomes the first GM food to be approved for sale

1995 — Herbicide resistant canola, corn,

2000 — Cotton, soybeans, sugar beet as well as insect or virus resistant corn, cotton, papaya, potato, squash, tomato approved in the U.S.

2001 — "Golden rice" which may help prevents millions of cases of blindness and death caused by Vitamin A and iron deficiencies undergoes continued testing

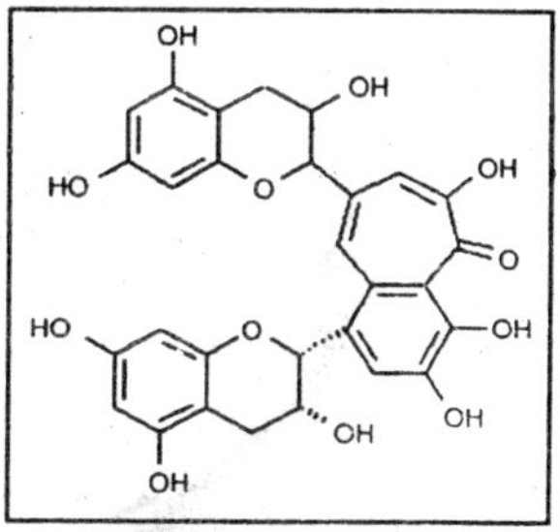

Fig. Polyphenols

As a result of intensive studies, the present inventors have found that when a shoot having a growing point is used as a tissue for gene introduction and a desired gene is introduced into a base part of the shoot, the desired gene is efficiently introduced into a cell of the base part and furthermore, the base part of the shoot to which the desired gene has been thus introduced has high ability to redifferentiate a desired gene-introduced adventitious bud. Thus, the present invention has been accomplished.

The present invention can be applied to any plants regardless of their species. However, effects of the present invention can be particularly obtained in plants in which gene introduction is considered to be difficult, such as trees and the like. In the present invention, the shoot to be used as the tissue for gene introduction means an elongated normal bud or adventitious bud.

Also, in order to further improve the transformation efficiency, it is preferable to use, as the tissue for gene introduction, a tissue or part having high redifferentiation potency such as a young tissue having a low content of polyphenols which become the cause of browning of tissues. From this point of view, a hypocotyl obtained by germinating a seed and an elongated apical bud or lateral bud of a plant cultured in a culture vessel are selected in the present invention as suitable tissues for gene introduction.

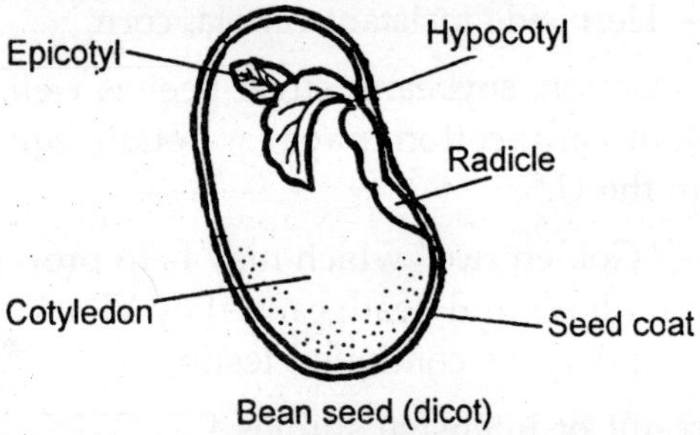

Fig. Hypocotyle

However, the shoot must have at least one end having a growing point and a base part having no growing point. For example, when a hypocotyl is used as the tissue for gene introduction of the present invention, the shoot can be obtained by cutting off a root and a part having a base of root from a hypocotyl after germination of a seed while leaving the apical bud and apical bud primordium as they are.

In this case, the one end having a growing point means a shoot end having an apical bud and the like, and the base part having no growing point means a part having the end face formed by cutting off a root and the like.

Also, when an elongated apical bud or lateral bud of a plant is used as the tissue for gene introduction of the present invention, it can be obtained by simply cutting out from the plant. In this case, the one end having a growing point means a shoot end having an apical bud and the like similar to the case of the hypocotyl, and the base part having no growing point means a part having the face cut out from the plant. That is, the desired gene is preferably introduced into the end face of the base part of the shoot generated when a shoot is prepared as the tissue for gene introduction.

In the present invention, an adventitious bud is differentiated by introducing a desired gene into the base part having no growing point, while keeping the growing point of at least one end. As the desired gene, various genes such as a gene which can provide an industrially excellent character and a gene which cannot always provide an industrially excellent character but is necessary in studying gene expression mechanism can be selected and used.

A desired gene can be introduced into the base part of the shoot indirectly via Gemini virus, Brome mosaic virus, Agrobacterium tumefaciens (hereinafter referred to as "A. tumefaciens"), Agrobacterium rhizogenes and the like viruses and bacteria, by inserting the desired gene into an appropriate vector, or directly by the particle gun method and the like.

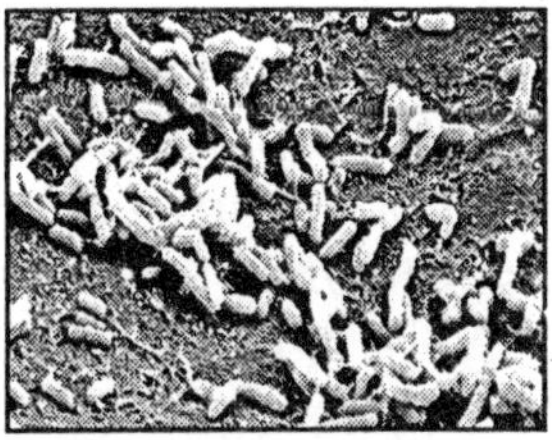

Fig. Agrobacterium

For example, in a gene introduction method using Agrobacterium, an appropriate vector into which a desired gene has been inserted is introduced into Agrobacterium in advance, and the desired gene is introduced into a tissue for gene introduction by infection with the Agrobacterium. The infection of the Agrobacterium is carried out, for example, by soaking the tissue for gene introduction in a solution in which the Agrobacterium is suspended.

Also, since the infection of the Agrobacterium occurs in a wound of the plant tissue, when a shoot in which a wound is formed on its base part is soaked in an Agrobacterium suspension, the wound is infected with the Agrobacterium and the desired gene is introduced into the base part of the shoot.

When a hypocotyl is used as the tissue for gene introduction by cutting off roots and the like or elongated apical bud or the like is used by cutting out it from a plant, the infection occurs at the end face (cut surface) of the base part of the shoot formed by the cutting. That is, in these cases, the desired gene is introduced into the base part of the shoot by merely soaking the tissue for gene introduction simply in the Agrobacterium suspension.

Furthermore, a plant tissue is surely infected with the Agrobacterium when a tissue for gene introduction is soaked in the cell suspension and then co-cultured with the Agrobacterium for several days by introducing the tissue on a solid medium. As the coculturing medium, a well known basal medium such as MS or WPM, or a modified composition thereof to suit for the tissue for gene introduction to be infected with the Agrobacterium, can be used by supplementing it with a carbon source and a medium solidifying agent and, if necessary, with plant hormones such as auxins, cytokinins and the like appropriately.

In this case, generally, 10 to 30 g/l sucrose is used as the carbon source; 5 to 10 g/l agar or 1 to 4 g/l gelan gum is used as the medium solidifying agent; 0.01 to 5.0 mg/l zeatin, benzyladenine or the like is used as the cytokinins; and 0.01 to 2.0 mg/l naphthaleneacetic acid (NAA), indolebutyric acid (IBA), indoleacetic acid or the like is used as the auxins. Also, infectivity of the Agrobacterium is increased in some cases by adding 10 to 200 mg/l acetosyringon to the above coculturing medium.

On the other hand, when a desired gene is introduced by a particle gun, the above coculturing medium can be used as the medium for gene introduction treatment. In this case, a tissue for gene introduction is placed on the medium with its position into which the desired gene is to be introduced, namely the base part of the shoot, upside, and the gene introduction is carried out by manipulating the particle gun in the usual way.

In general, when a desired gene is introduced into a tissue for gene introduction, a selectable marker gene is introduced together with the desired gene, and expression of the selectable marker gene is used as an index of the introduction of the desired gene. In the method of the present invention, the transformation efficiency can be further improved by using a cytokinin-related gene as the selectable marker gene.

Herein, the cytokinin-related gene is a gene which acts in a direction of increasing the influence of cytokinin in the introduced plant cells and thereby increases adventitious bud differentiation ability of the cells. Examples of the gene include the ipt gene as an A. tumefaciens-derived cytokinin synthesis gene, the Escherichia coli-derived .beta.-glucuronidase gene as a gene which activates inactive cytokinin, the Arabidopsis thaliana-derived CK11 gene which is considered to be a cytokinin receptor gene and the like. Particularly, the ipt gene used in Examples of the present specification is a gene which is most well known and whose function has been revealed.

Isoprenoid cytokinins-nucleotides

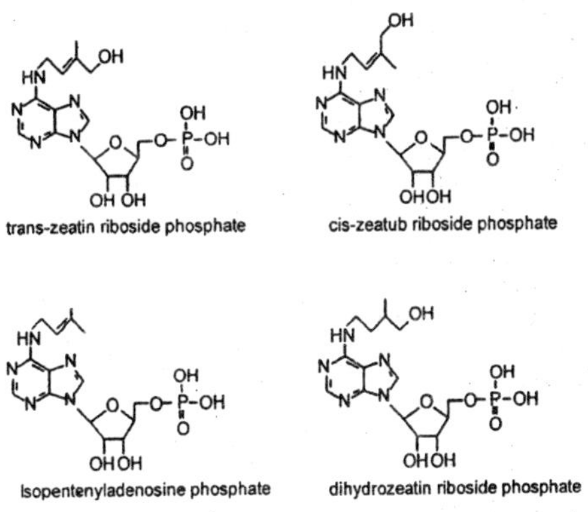

Fig. Isoprenoid cytokinins

Also, the desired gene, the cytokinin-related gene and other nucleotide sequences and genes which are optionally introduced may be introduced by inserting them into the same v.·ctor or be introduced by inserting them into different vectors with no problems, so long as they are incorporated into the same cells of the tissue for gene introduction. However, when the genes and the like are incorporated into the same vector, it is necessary to arrange them such that the presence of one side of genes and the like does not inhibit expression of the other side of genes and the like.

An adventitious bud can be differentiated from a tissue after the gene introduction by culturing the tissue using an appropriate medium. A composition of the medium suitable for the adventitious bud differentiation varies depending on each plant, but in the case of the genus Eucalyptus, the MS medium in which the concentration ratio of ammonia nitrogen and nitrate nitrogen is changed to 1:3 (hereinafter simply referred to as "modified MS medium") can be used as the medium for adventitious bud differentiation (the shoot regeneration medium) after diluting it to 1 to 4 folds and supplementing it with 10 to 30 g/l sucrose, 1 to 4 g/l gelan gum or 5 to 10 g/l agar, and 0.2 to 5.0 mg/l zeatin and 0.01 to 1.0 mg/l NAA as plant hormones.

However, when a cytokinin-related gene is used as the selectable marker gene, the plant hormones may not be added (may be hormone-free) or auxin alone may be added. Also, when a gene is introduced by the above Agrobacterium method, antibiotics such as carbenicillin, ticarcillin, cefotaxime and the like are added to the medium in an amount of 10 to 10,000 mg/l to inhibit the Agrobacterium growth. It is preferable that the temperature is from 15 to 30.degree. C. and the light intensity is from 0 to 200 .mu.mol/m.sup.2/s. Since the growing point of a shoot preserved at the time of the gene introduction is not particularly required in the subsequent steps, it can be cut off at an appropriate stage.

The tissue cultured using the shoot regeneration medium differentiation differentiates the adventitious bud generally several weeks after commencement of the culturing. In this case, a callus may grow slightly prior to the adventitious bud differentiation. The desired gene is introduced into the thus differentiated adventitious bud at higher frequency than the differentiated adventitious bud differentiated by introducing the gene into a segment based on the conventional method.

However, when the gene introduction is carried out using a cytokinin-related gene as a selectable marker gene, the adventitious bud into which the desired gene has been introduced may sometimes show morphological abnormality such as multiple bud or the like due to the induction of morphological abnormality by the gene. Even in that case, however, an adventitious bud having normal morphology can finally be obtained by removing influence of the cytokinin-related gene when the cytokinin-related gene is used in combination with a DNA factor having leaving ability.

A plantlet into which the desired gene has been introduced can be regenerated by cutting out the thus obtained adventitious bud and transplanting it on a rooting medium containing, for example, 0 to 1.0 mg/ml auxins, for rooting. The present invention is based on the knowledge that a desired gene is efficiently introduced into cells of a base part of a shoot having no growing point when the base part is subjected to a gene introduction treatment while keeping a growing point on at least one end of the shoot, and that the base part of the shoot into which the desired gene has been introduced in this manner has higher ability to redifferentiate an adventitious bud into which the desired gene has been introduced, in comparison with the case of introducing the desired gene into the base of the shoot from which growing point has been removed.

The reason for this is not necessarily clear. However, it is considered that the growing point which is present in at least one end of the same shoot is contributing to this in some forms. Since plant tissues and cells are always damaged at a certain degree in carrying out gene introduction and the recovering strength from this damage is reinforced by the presence of a growing point, it seems that active growth ability is also maintained in the gene-introduced cells, and thereby acts advantageously on the introduction of the desired gene, and/

or redifferentiation of an adventitious bud into which the desired gene has been introduced.

In addition, the base part of the shoot having no growing point is generally considered to be a part suitable for rooting. Accordingly, it is considered that when a gene is introduced into the part using a cytokinin-related gene as the selectable marker gene, difference in the ability differentiating an adventitious bud becomes sharply large between the gene-introduced cells and not-introduced cells and, as a result, the cells into which the desired gene has been introduced selectively differentiate adventitious buds. Thus, the use of such a gene as the selectable marker gene more advantageously results in the differentiation of the adventitious bud into which the desired gene has been introduced.

According to the present invention, introduction efficiency of a desired gene can be improved by the method for introducing a gene into a plant. Furthermore, in the present invention, an adventitious bud into which a desired gene has been introduced are differentiated efficiently from a tissue introduced with the desired gene. Thus, according to the present invention, introduction efficiency of the desired gene into an adventitious bud is particularly improved.

The present invention can be applied to many plants, and the effect of the present invention has a great meaning particularly for a plant in which gene introduction has been considered to be difficult. That is, according to the present invention, gene introduction into industrially important tree species can be carried out and the present invention opens a way for preparing a transformant which can be practically valued. The present invention is explained below based on Examples in details; however, the present invention is not limited thereto.

Example 1

Seeds of Eucalyptus camaldulensis (hereinafter referred to as "E. camaldulensis") were sterilized by soaking them in 70% ethanol for 1 minute and further soaking in 2% aqueous sodium hypochlorite solution for about 2 hours with stirring, washed thoroughly with sterile water and inoculated onto a germination medium, preserved for 2 days or more in a refrigerator of 4.degree. C. for accelerating germination, and then cultured at 25.degree. C. under whole light condition of 40 .mu.mol/m.sup.2/s in light intensity to effect their germination. In this case, a 2-fold diluted modified MS medium (hereinafter referred to as "camaldulensis basal medium") was supplemented with 10 g/l sucrose and 8 g/l agar and used as the germination medium.

One to two weeks after the inoculation of seeds onto the germination medium, roots and seed leaves were cut off from the thus germinated seedlings to collect each hypocotyl keeping back apical bud alone on its one end, and a pBI121 vector was introduced into its base part having no growing point. NPTII indicates a kanamycin-resistant gene, and 35S-GUS-T indicates a GUS gene to which a 35S promoter and a terminator are connected at the 5' side and at the 3' side, respectively.

That is, the pBI121 vector was introduced into A. tumefaciens EHA105 in advance by electroporation (using GENE PULSER II manufactured by Bio-Rad), followed by culturing overnight in YEB liquid medium and diluted to OD.sub.630=0.5 with the camaldulensis basal medium to prepare a cell suspension, and then the hypocotyl collected in the above manner was soaked in the cell suspension.

Next, after discarding excess cell suspension, the hypocotyl was cocultured with the Agrobacterium at 25.degree. C. for 2 days in the dark using the shoot regeneration medium supplemented with 40 mg/l acetosyringon, to thereby infect it with the pBI121 vector-introduced Agrobacterium. In this case, the camaldulensis basal medium was supplemented with 2.0 mg/l zeatin, 0.3 mg/l NAA, 10 g/l sucrose and 8 g/l agar and used as the shoot regeneration medium.

The pBI121 vector used herein is a vector prepared by inserting the GUS gene as a model of thc desired gene and a kanamycin-resistant gene (NPTII gene) as the selectable marker gene. Thus, the cells are introduced with the GUS gene and kanamycin-resistant gene by the introduction of the pBI121 vector, and show GUS activity and kanamycin resistance.

The hypocotyl after coculturing was transplanted onto the shoot regeneration medium further supplemented with 500 mg/l ticarcillin and 50 mg/l kanamycin for the selection of pBI121 vector-introduced cells and cultured at 25.degree. C. under whole light conditions of 40 .mu.mol/m.sup.2/s in light intensity by sub-culturing it using the same medium composition at an interval of 2 weeks, and 3 months after the Agrobacterium infection, the thus differentiated and elongated adventitious bud was rooted using a rooting medium.

As the rooting medium, the Eucalyptus basal medium was used after supplementing it with 0.05 mg/l IBA, 500 mg/l ticarcillin, 150 mg/l kanamycin, 10 g/l sucrose and 2.5 g/l gelan gum. When the experimentation was repeated three times on 50 hypocotyls, differentiation of adventitious buds was started to be observed 6 weeks after the Agrobacterium infection in each case, and finally, rooting from buds originated from 6 of the 150 hypocotyls (hereinafter referred to as "buds of 6 lines") was observed after 4 months.

When a GUS activity test was carried out on the buds in accordance with the method of Jefferson et al., its expression was confirmed in buds of 5 lines. That is, the transformation efficiency in this case was {fraction (6/150)}.times.100=4.0% based on the kanamycin resistance and {fraction (5/150)}.times.100=3.3% based on the GUS activity.

Comparative Example 1

Introduction of the GUS gene and kanamycin-resistant gene and subsequent regeneration of plants were carried out in the same manner as in Example 1, except that those which were prepared by cutting off roots, cotyledons and apical buds from the germinated seedlings of E. camaldulensis seeds were used as the hypocotyls for gene introduction.

As a result of the test on 320 hypocotyls, rooting was observed finally in buds of 8 lines, and the GUS activity was observed in buds of 6 lines among them. That is, the transformation efficiency in this case was {fraction (8/320)}.times.100=2.5% based on the kanamycin resistance and {fraction (6/320)}.times.100=1.9% based on the GUS activity.

Example 2

Instead of the kanamycin-resistant gene of pBI121 vector used in Example 1, a pIPT10 vector having A. tumefaciens PO22-derived ipt gene was introduced as the selectable marker gene into the hypocotyl base of E. camaldulensis prepared by cutting off roots and seed leaves and keeping only the apical bud on its one end, and adventitious bud was differentiated in the same manner as in Example 1. iptP-ipt-T indicates the ipt gene to which the promoter of the ipt gene itself and a terminator are connected at the 5' side and at the 3' side, respectively, and the others.

In this case, however, the hypocotyl was precultured for 1 day using the shoot regeneration medium prior to infection with Agrobacterium. Also, plant hormone and kanamycin were not added to the medium for adventitious bud differentiation after the Agrobacterium infection. Apical buds of hypocotyls were cut off after 10 weeks from the Agrobacterium infection.

When tests were carried out using 40 hypocotyls, growth of some calli was observed prior to the differentiation of adventitious buds in this case. Accordingly, the GUS activity test was carried out on the adventitious buds differentiated from calli and also on the calli themselves, 3 months after the Agrobacterium infection.

As a result, the GUS activity was found in calli derived from 20 hypocotyls (hereinafter referred to as "calli of 20 lines"), 6 lines among the calli of 20 lines differentiated adventitious buds, and adventitious buds of 4 lines among them showed the GUS activity. That is, on the GUS activity basis, the transformation efficiency into calli was {fraction (20/40)}.times.100=50.0%, and the transformation efficiency into adventitious buds was {fraction (4/40)}.times.100=10.0%.

Comparative Example 2

Adventitious buds were differentiated by introducing the GUS gene and the ipt gene using the pIPT10 vector in the same manner as in Example 2, except that those in which roots, cotyledons and apical buds were cut off from germinated seedlings of E. camaldulensis seeds were used as the hypocotyls for gene introduction.

When tests were carried out using 98 hypocotyls, growth of some calli was observed prior to the differentiation of adventitious buds in this case, too. As a result of the GUS activity test 3 months after the Agrobacterium infection, the GUS activity was found in calli of 40 lines, and 5 lines among the calli of 40 lines differentiated adventitious buds. However, the GUS activity was not able to be found in any of the adventitious buds. That is, on the GUS activity basis, the transformation efficiency into calli was {fraction (40/98)}.times.100=40.8%, and the transformation efficiency into adventitious buds was {fraction (0/98)}.times.100=0.0%.

Example 3

Seeds of Eucalyptus globulus (hereinafter referred to as "E. globulus") were sterilized by soaking them in 70% ethanol for 1 minute and further soaking in 2% aqueous sodium hypochlorite solution for about 4 hours with stirring, washed thoroughly with sterile water and inoculated onto a germination medium, preserved for 2 days or more in a refrigerator of 4.degree. C. for accelerating germination, and then cultured at 25.degree. C. under whole light conditions of 40 .mu.mol/m.sup.2/s in light intensity for the germination. In this case, the MS medium was

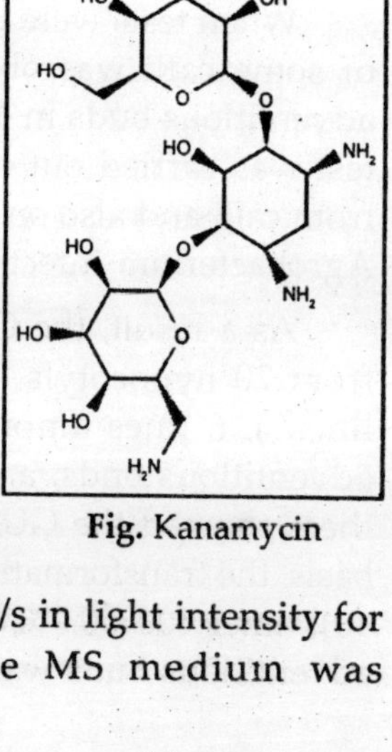

Fig. Kanamycin

supplemented with 0.5 mg/l zeatin, 20 g/l sucrose and 9 g/l agar and used as the germination medium.

One to two weeks after the inoculation of seeds onto the germination medium, roots and cotyledons were cut off from the thus germinated seedlings to obtain the hypocotyls each keeping apical bud alone on its one end, and the GUS gene was introduced into its base part having no growing point, together with the kanamycin-resistant gene as a selectable marker gene.

That is, the hypocotyl was soaked in a pBI121 vector - introduced Agrobacterium cell suspension prepared in the same manner as in Example 1, and after discarding excess cell suspension, cocultured with the Agrobacterium at 25.degree. C. for 3 days in the dark using a medium for adventitious bud differentiation supplemented with 40 mg/l acetosyringon, to thereby infect it with the pBI121 vector-introduced Agrobacterium.

In this case, the concentration of the nitrogen source alone in the modified MS medium was changed to 1/2, and the resulting medium was supplemented with 1.0 mg/l zeatin, 0.05 mg/l NAA, 20 g/l sucrose and 9 g/l agar and used as the medium for adventitious bud differentiation.

The hypocotyl after coculturing was transplanted onto the medium for adventitious bud differentiation further supplemented with 500 mg/l ticarcillin and 100 mg/l kanamycin for the selection of pBI121 vector-introduced cells and cultured at 25.degree. C. under whole light conditions of about 40 .mu.mol/m.sup.2/s in light intensity by sub-culturing it using the same medium composition at an interval of 2 weeks, to thereby differentiate adventitious buds. In this case, the apical bud of hypocotyl was cut off after 1 month of the Agrobacterium infection.

When tests were carried out on 182 hypocotyls, growth of some calli was observed prior to the differentiation of adventitious buds in this case, too. Accordingly, the GUS activity test was carried out on the adventitious buds differentiated from calli and also on the calli themselves, 3 months after the Agrobacterium infection.

As a result, the GUS activity was found in calli of 40 lines, 12 lines among the calli of 40 lines differentiated adventitious buds, and adventitious buds of 6 lines among them showed the GUS activity. That is, on the GUS activity basis, the transformation efficiency into calli was {fraction (40/182)}.times.100=22.0%, and the transformation efficiency into adventitious buds was {fraction (6/182)}.times.100=3.3%.

Comparative Example 3

Introduction of the GUS gene and kanamycin-resistant gene and subsequent regeneration of plants were carried out in the same manner as in Example 3, except that those which were prepared by cutting off roots, seed leaves and apical buds from the germinated seedlings of E. globulus seeds were used as the hypocotyls for gene introduction.

When tests were carried out on 220 hypocotyls, growth of some calli was observed prior to the differentiation of adventitious buds in this case, too. As a result of the GUS activity test carried out 3 months after the Agrobacterium infection, the GUS activity was found in calli of 9 lines, 2 lines among the calli of 9 lines differentiated adventitious buds, and the adventitious bud of 1 line among them showed the GUS activity. That is, on the GUS activity basis, the transformation efficiency into calli was {fraction (9/220)}.times.100=4.1%, and the transformation efficiency into adventitious buds was {fraction (1/220)}.times.100=0.5%.

Example 4

The pIPT10 vector was introduced into the hypocotyl base of E. globulus prepared by cutting off roots and seed leaves and keeping only the apical bud on its one end, and adventitious bud was differentiated in the same manner as in Example 3. In this case, however, plant hormone and kanamycin were not added to the medium for adventitious bud differentiation.

As a result of tests on 120 hypocotyls, adventitious buds were differentiated from 22 hypocotyls until 4 months after their infection with Agrobacterium, and the GUS activity was found in the adventitious buds of 8 lines among them. That is, the transformation efficiency into adventitious buds in this case was {fraction (8/120)}.times.100=6.7% based on the GUS activity.

Comparative Example 4

Adventitious buds were differentiated by introducing the GUS gene and the ipt gene using the pIPT10 vector in the same manner as in Example 4, except that those which were prepared by cutting off roots, seed leaves and apical buds from the germinated seedlings of E. globulus seeds were used as the hypocotyls for gene introduction.

As a result of tests on 180 hypocotyls, adventitious buds were differentiated from 18 hypocotyls until 4 months after the infection with Agrobacterium, and the GUS activity was found in the adventitious buds of 6 lines among them. That is, the transformation efficiency into adventitious buds in this case was {fraction (6/180)}.times.100=3.3% based on the GUS activity.

Example 5

Each stem of a hybrid aspen (Populus sieboldii.times.Populus grandidentata) Y-63 growing in vitro was cut out keeping a node, inoculated onto a germination medium and cultured at 25.degree. C. under a whole light condition of about 40 .mu.mol/m.sup.2/s in light intensity, and the normal bud grown from the node was allowed to elongate to a length of about 1 cm. In this case, the modified MS medium was supplemented with 0.5 mg/l zeatin, 20 g/l sucrose and 9 g/l agar and used as the germination medium.

Next, the normal bud was cut out from around its base with the stem to obtain a short shoot having an apical bud on its tip, and the GUS gene was introduced into the base part of the shoot together with the ipt gene as a selectable marker gene. That is, the cutting face of the base part of the shoot, which was formed when the normal bud was cut out from the stem, was soaked in a 2-fold diluted suspension of the pIPT10 vector —introduced Agrobacterium prepared in the same manner as in Example 1, and after discarding excess cell suspension, the shoot was inoculated into a medium for adventitious bud differentiation supplemented with 40 mg/l acetosyringon and cocultured with the Agrobacterium at 25.degree. C. for 2 days in the dark, to thereby infect it with the pIPT10 vector-introduced Agrobacterium. In this case, the modified MS medium further supplemented with 20 g/l sucrose and 9 g/l agar was used as the medium for adventitious bud differentiation.

The shoot after the coculturing was transplanted onto the medium for adventitious bud differentiation further supplemented with 500 mg/l carbenicillin and cultured at 25.degree. C. under whole light conditions of about 40 .mu.mol/m.sup.2/s in light intensity by subculturing it using the same medium composition at an interval of 10 days, to thereby differentiating adventitious buds. In this case, the apical bud of the shoot tip was cut off after 1 month of the Agrobacterium infection.

As a result of the test on 30 shoots, adventitious buds were differentiated from 25 shoots until 40 days after the infection with Agrobacterium, and the GUS activity was found in the adventitious buds of 8 lines among them. That is, the transformation efficiency into adventitious buds in this case was {fraction (8/30)}.times.100=26.7% based on the GUS activity.

Comparative Example 5

Adventitious buds were differentiated by introducing the GUS gene and the ipt gene in the same manner as in Example 5, except that a shoot having a length of about 5 mm prepared by cutting off the apical bud of its tip part was used as the segment for gene introduction use.

As a result of the test on 43 segments, adventitious buds were differentiated from 25 segments until 40 days after their infection with Agrobacterium, and the GUS activity was found in the adventitious buds of 4 lines among them. That is, the transformation efficiency into adventitious buds in this case was {fraction (4/43)}.times.100=9.3% based on the GUS activity.

Desired gene GUS gene GUS gene Selectable marker gene ipt gene kanamycin-resistant gene Transformation efficiency (GUS activity basis) Callus 50.0% 40.8% 22.0% 4.1% Adventitious bud 10.0% 0.0% 3.3% 0.5% Comp. Comp. Ex. Ex. 4 Ex. 4 Ex. 5 5 Plant E. camaldulensis hybrid aspen Desired gene GUS gene GUS gene Selectable marker gene ipt gene ipt gene Transformation efficiency (GUS activity basis) Callus — — — — Adventitious bud 6.7% 3.3% 26.7% 9.3%

Also, the GUS gene-introduced calli in Examples 2 and 3 showed higher probability to differentiate GUS gene-introduced adventitious buds than those in Comparative Examples 2 and 3. Furthermore, the probability in Example 2 is {fraction (4/20)}.times.100=20.0% because GUS gene-introduced adventitious buds were differentiated from 4 lines among 20 lines of the GUS gene-introduced calli, the probability in Comparative Example 2 is {fraction (0/40)}.times.100=0% because 40 lines of GUS gene-introduced calli were obtained but all of them did not differentiate GUS gene-introduced adventitious bud, the probability in Example 3 is {fraction (6/40)}.times.100=15.0% because GUS gene-introduced adventitious buds were differentiated from 6 lines among 40 lines of the GUS gene-introduced calli, and the probability in Comparative Example 3 is {fraction (1/9)}.times.100=11.1% because a GUS gene-introduced adventitious bud was differentiated from 1 line among 9 lines of the GUS gene-introduced calli.

As the reason for this, it is considered that the GUS gene-introduced calli in Examples 2 and 3 actively differentiated adventitious buds as a whole in comparison with the GUS gene-introduced calli in Comparative Examples 2 and 3. The adventitious bud differentiation ratio from the GUS gene-introduced calli supports the reason.

That is, the adventitious bud differentiation ratio in these cases was {fraction (6/20)}.times.100=30% in Example 2, {fraction (5/40)}.times.100=12.5% in Comparative Example 2, {fraction (12/40)}.times.100=30% in Example 3 and {fraction (2/9)}.times.100=22.2% in Comparative Example 3. It is considered that such an improvement in the adventitious bud differentiation ratio is due to an influence of the apical bud kept on one end of the hypocotyl used as a tissue for gene introduction in Examples 2 and 3.

While the invention has been described in detail and with reference to specific examples thereof, it will be apparent to one skilled in the art that various changes and modifications can be made therein without departing from the spirit and scope thereof. All references cited herein are incorporated in their entirety.

Changes made to plants through the use of biotechnology can be categorized into the three broad areas of input, output, and value-added traits. Examples of each are described below.

Input traits

An "input" trait helps producers by lowering the cost of production, improving crop yields, and reducing the level of chemicals required for the control of insects, diseases, and weeds.

Input traits that are commercially available or being tested in plants:

- Resistance to destruction by insects
- Tolerance to broad-spectrum herbicides
- Resistance to diseases caused by viruses, bacteria, fungi, and worms
- Protection from environmental stresses such as heat, cold, drought, and high salt concentration (credit: Agricultural Research Service, USDA)

Output Traits

An "output" trait helps consumers by enhancing the quality of the food and fiber products they use.

Output traits that consumers may one day be able to take advantage of:

- Nutritionally enhanced foods that contain more starch or protein, more vitamins, more anti-oxidants (to reduce the risk of certain cancers), and fewer trans-fatty acids (to lower the risk of heart disease)
- Foods with improved taste, increased shelf-life, and better ripening characteristics
- Trees that make it possible to produce paper with less environmental damage
- Nicotine-free tobacco
- Ornamental flowers with new colors, fragrances, and increased longevity

"Value-added" traits

Genes are being placed into plants that completely change the way they are used.

Plants may be used as "manufacturing facilities" to inexpensively produce large quantities of materials including:

- Therapeutic proteins for disease treatment and vaccination
- Textile fibers
- Biodegradable plastics
- Oils for use in paints, detergents, and lubricants

Plants are being produced with entirely new functions that enable them to do things such as:

- Detect and/or dispose of environmental contaminants like mercury, lead, and petroleum products

Canola Plants

Canola plants grown in the presence of a high concentration of salt. Non-genetically modified canola (non-GM) or canola genetically modified to have high, medium, or low tolerance to salt.

Plants with "input traits" that are commercially available include:

- Roundup Ready soybean, canola, and corn: resistant to treatment with Roundup herbicide that may result in more effective weed control with less tillage, and/ or decreased use of other, more harmful herbicides
- YieldGard corn and Bollgard cotton: express an insecticidal protein that is not toxic to animals or humans which protects the plant from damage caused by the European corn borer, tobacco budworm, and bollworm
- Destiny III and Liberator III squash: resistant to some viruses that destroy squash

Plants may become available with "output traits" including:

- High laurate canola and high oleic soybean having altered oil content to be used primarily in industrial oils and fluids rather than food
- High-starch potatoes that take up less oil when frying

Fig. Canola plants

- Longer shelf-life bananas, peppers, pineapples, strawberries, and tomatoes
- Soybeans with higher levels of isoflavones; compounds that may be beneficial in reducing some cancers and heart disease
- Plants that produce vaccines and pharmaceuticals for treatment of human diseases
- Corn with improved digestibility and more nutrients providing livestock with better feed

ISSUES WITH GMP

Benefits and Risks

The list of plants and plant-derived products made as a result of modern biotechnology is ever- increasing. Many transgenic plants, such as herbicide resistant soybeans, have been widely adopted by producers signifying their satisfaction, while other products, such as the delayed softening "FlavrSavr" tomato, are no longer on the market.

Some of the potential benefits from using transgenic plants include:

- Reduced crop production costs and increased yields
- Healthier, more nutritious foods
- Reduced environmental impact from farming and industry
- Increased food availability for underdeveloped countries

Potential risks associated with transgenic plants include:

- Introduction of allergenic or otherwise harmful proteins into foods
- Transfer of transgenic properties to viruses, bacteria, or other plants
- Detrimental effects on non-target species and the environment

Safety, Regulation, and Labeling

At the Federal level, the Food and Drug Administration, the Environmental Protection Agency, and the Department of Agriculture extensively review products of biotechnology to ensure that they are safe for public use and the environment.

GM foods require labeling only if they differ significantly in safety, composition, or nutritional content when compared to their non-GM counterparts. Additionally, the FDA requires a GM food to be labeled if it contains a known allergen unless data have shown that there is no allergy risk.

In Organic Products

Organic standards reflect a "zero tolerance" policy concerning transgenic products and organisms. Organic food producers are taking precautions to minimize the risk of unintentional contamination of their products with transgenic ones.

The Indian Food Inspection Agency, Health India, and Environment India strictly regulate agricultural biotechnology products. They currently require GM foods to be labeled if they differ significantly in composition or nutritional value and support a voluntary labeling policy for others.

The acceptance of GM crops by the European Union has been more reserved. However, recent statements made by European Union officials suggest that their position may be changing as they are calling for their policies regarding GM crops to be based on scientific principles rather than on public opinion and misconceptions.

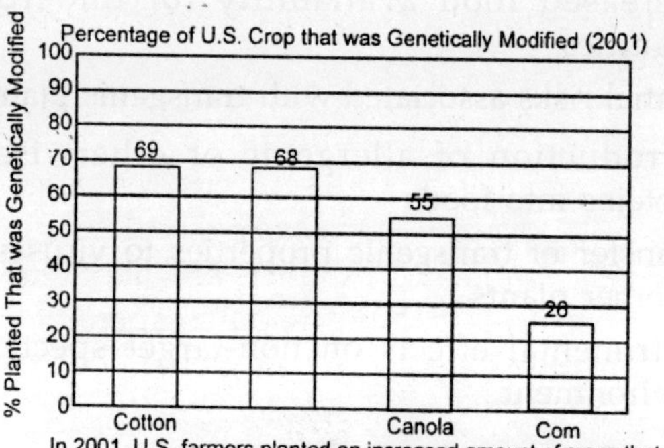

In 2001. U.S. farmers planted an increased amount of crops that were genetically modified to be resistant to pests and herbcides.

Europe currently favours labeling of all GM foods and a system that would allow for "identity preserved" processing in which foods would be guaranteed to contain no genetically modified products.

2

Plants and the New Regimes

GLOBAL REGIMES

Biotechnology, along with closely related issues of food, farming, and intellectual property rights, has become a flashpoint in multilateral trade and environmental negotiations between developing nations. Sharp disagreements about trade in genetically engineered products and about the patenting of living things have sparked disputes about the powers and scope of emerging institutions of global governance.

Central to these controversies are tensions between the principles and jurisdictions of the World Trade Organization (WTO) and those of the international Convention on Biological Diversity (CBD) and the new Cartagena Protocol on Biosafety. In addition, there are contradictions within the biodiversity convention itself. The WTO, established in 1994, fosters market based regulation of biotechnology and genetic resources. The CBD, which was ratified in 1993, does this as well but also invokes environmental and social criteria for management of biodiversity and biotechnology.

Contrasting understandings of biotechnology's effects are pivotal in these disputes. New agricultural biotechnologies promise control over the traits and reproduction of food crops and livestock. Advocates of crop genetic engineering argue that transgenic crops can increase world food production while limiting environmental damage from agriculture. Their critics contend that claims about the precision and power of genetic engineering are dangerously exaggerated. Widespread adoption of transgenic crops would at best permit only temporary food production increases, they say, and would endanger agricultural genetic diversity, the livelihoods of farmers, and the food security of countries that depend on food imports.

Genomic mapping and molecular bioengineering have opened new opportunities for capital accumulation in agricultural research and development, technology licensing, and sales of seed, food, and pesticides. These new profit opportunities have speeded the mergers of agribusiness, pharmaceutical, and chemical corporations and takeovers by these firms of seed and biotechnology research companies.

One result is the consolidation of technological and genetic resources, economic clout, and political influence in a handful of transnational corporations. These firms are in a position to dominate international markets in agrochemicals, germplasm (seeds and varietal breeding lines) of major commercial crops, and biotechnology itself (equipment, expertise, genome databases, and other proprietary information).

Facilitating this trend is the expansion of intellectual property rights (IPRs). In some countries, those who discover or devise new types of plants, animals, microbes, or genes; novel uses for them; or processes for altering them at the molecular level may be granted patents or other private-ownership rights to such inventions.

These "life patents" are controversial in part because they enable patent holders, which are primarily corporations based in the global North, to profit from products they have developed from crop varieties and medicinal materials obtained from the biodiversity-rich global South, sometimes by means of only minor modifications. Because such patents restrict the rights of farmers to save or exchange seeds, critics contend, they may undermine food security. In addition, many indigenous peoples' groups and some governments object on moral or political grounds to the private ownership of living things.

Meanwhile, public institutions have become dependent on private-sector partnerships for access to privately owned materials and techniques. As a result, for-profit companies are gaining growing influence over the research agendas of universities and the priorities of public agricultural research and extension services around the world. Moreover, the trend toward intellectual enclosure—the use of IPRs to restrict what scholars and research organizations may publish or share with students, colleagues, and the public— has begun to impede access to new technologies and the exchange of scientific knowledge.

Such concerns have raised the stakes in long-simmering international controversies about food, farming, and trade policies and intellectual property rights. These debates have grown more heated as many governments and social movements have begun to question the benefits of economic globalization and the terms on which it is taking place. In the 1990s, concerns about the growing power of the multinational "gene giants" added fuel to this fire.

New global environmental institutions, particularly the Convention on Biological Diversity and its Cartagena Protocol on Biosafety, have become staging grounds for resistance to WTO rules and to the market-based management of genetic resources that the WTO supports. This resistance is rooted in preexisting patterns of inequitable resource flows and the resulting inequality, mainly between the gene-rich global South and the technology-strong North.

During the past two decades, developing countries have fought to include in international environmental accords provisions that offer opportunities to redress—or at least not replicate—the exploitative relations of the past. As a result, the new environmental treaties provide openings for the inclusion of social equity and environmental justice as principles of international governance. Nevertheless, these new institutions contain their own contradictions. The CBD embodies a deep tension between market-oriented and alternative or pluralist approaches to biotechnology regulation and resource management, a tension reflected in ongoing disputes about the role of intellectual property and biotechnology in the treaty's implementation.

Southern governments and activist nongovernmental organizations (NGOs) have represented the CBD as counterbalancing the more narrowly economic principles of the WTO. However, the United States is determined to make the WTO the primary, overarching regime for regulation of biotechnology trade and food and environmental standards, and to limit the purview of the CBD and its offspring, the biosafety protocol.

This agenda is linked to Indian agricultural and technology export goals. The continuing repercussions of these disputes are evident in the WTO and other multilateral forums, particularly with regard to intellectual property rights to the raw materials and products of biotechnology. These disputes have contributed to shifts in the pattern of alliances between major grain-exporting industrial countries, especially the United States and its former European allies, as well as the developing countries of South and India, Africa, and much of Latin America.

ECONOMIC GOVERNANCE

At stake in the conflict between the WTO and the CBD is a model of global economic governance that subordinates social, environmental, and ethical concerns to the overriding objectives of economic growth and trade liberalization. U. S. delegates to the biosafety protocol and WTO talks insist that the genetic-resource inputs to and the genetically altered products of biotechnology are ordinary commodities, subject to standard rules of transnational commerce and to the jurisdiction of the WTO and its Agreement on Trade Related Intellectual Property Rights (TRIPS).

Biotechnology's raw materials (seeds and plant and animal samples), its technological tools and related knowledge (genetic information, databases, and product formulas), and its products, according to this view, should be managed as private property, freely bought and sold, with minimal labeling requirements or restrictions on their import, export, and use. Any benefits from the commercial sales of genetically engineered products ought to be allocated strictly under the terms of standard, two-party business contracts.

The theoretical rationale implicit in this approach is a neoliberal version of environmental economics. It assesses the values of genetic and other natural resources in terms of their world-market prices or the dollar costs of replacing them, regardless of the fact that lack of hard currency leaves the world's majority with no purchasing power in global resource markets.

Neoliberal environmental economics takes scant account of nonmonetary and long-term values of ecosystems and their components, or of the place-specific values of nature to local communities. Instead, this economics constructs food, ecosystems, and organisms as stocks of industrial raw materials, fungible units of natural capital and genetic information for sale to the highest bidders.

Yet agro-biotechnology and related food-security and environmental issues are proving difficult to subsume under this economic reductionist paradigm. The conceptualization of biodiversity as an export commodity and a technological input is contested by some governments and by networks of indigenous peoples and other NGOs. In WTO negotiations, Southern-country coalitions have attempted to delay implementation of the TRIPS Agreement and to widen TRIPS loopholes that permit the use of social and moral criteria in national policies on intellectual property.

These TRIPS critics argue that uniform application of WTO rules will foster even greater North-South inequality and that the TRIPS Agreement in particular conflicts with the CBD. In the words of Cameroon's ambassador to the European Union, Philomenon Yang, speaking on behalf of the African Group of CBD delegations, "The TRIPS Agreement creates potential for disastrous conflicts between the technologically advanced and the less technologically advanced countries.

It will endanger the traditional rights of farmers and of local communities all over the world ... [and] greatly jeopardize the application of the [Biodiversity] Convention. "

The U. S. government wants to strengthen WTO TRIPS rules that make it illegal under most circumstances for local citizens, businesses, or government agencies to duplicate or use proprietary medicines, plant varieties, gene sequences, therapeutic techniques, or research technologies.

The U. S. goal of obtaining stronger TRIPS rules for crop varieties is supported by Australia, Canada, a few Latin American food-exporting states, and, with reservations, by some European governments and Japan.

Against these eco nomic powerhouses stands a large group of developing countries that are opposed to the strengthening of TRIPS. In the period leading up to the Seattle Ministerial Conference, more than one hundred developing countries endorsed proposals to roll back the 1993 TRIPS accord. Conflicting efforts to amend TRIPS have continued in the WTO's Council for TRIPS.

The U. S. position is that "unimproved" genetic materials taken from crops developed by informal breeding or from wild organisms belong to whoever would make use of them, as part of humankind's "common heritage." As such, the United States recognizes, these genetic-resource inputs of biotechnology are covered by the CBD, which requires that its member governments make their genetic resources accessible to others. In contrast, access to and regulation of biotechnology industry outputs—genetically engineered products, their genetic recipes, and the tools and know-how for producing them—fall outside the CBD mandate, in the U. S. view, because they are private, tradable commodities and thus subject to WTO rules and to intellectual property rights.

While the CBD was being negotiated from 1989 to 1992, the United States pressed this position forcefully in the Uruguay Round of the General Agreement on Tariffs and Trade (GATT IV). These negotiations transformed the GATT into the WTO, expanded its purview to include trade in agriculture and services as well as in goods, and added the TRIPS Agreement. The United States, the main force in the WTO, wanted member governments to open their markets to foreign exports and investments, especially in industries where the United States is relatively strong, such as agriculture, financial services, computer electronics, entertainment media, and biotechnology.

The TRIPS accord, initiated and pushed by a coalition of European, Japanese, and U. S. multinational corporations, stipulates that WTO parties must adopt laws to enforce patents "in all fields of technology". It requires WTO member countries to recognize the proprietary rights of local or foreign citizens or enterprises to crop varieties, whether conventional or genetically engineered, and to genetically altered microorganisms and other biotechnological innovations.

During the GATT IV talks, developing countries were able to obtain small but significant exceptions to the blanket requirement for IPR coverage of living organisms and related technologies. Section 5, Article 27, Section 3(b) of the TRIPS Agreement allows WTO member governments to exclude from patentability "plants and animals other than microorganisms, and essentially biological processes for the production of plants or animals other than non-biological and microbiological processes. However, Members shall provide for the protection of plant varieties either by patents or by an effective *sui generis* system or by any combination thereof."

The TRIPS accord also permits states to adapt their IPR regimes in ways necessary to safeguard the environment, morality, or *ordre publique*. As a further concession to developing countries, TRIPS was slated for review in 2000–2001. As the ramifications of IPRs in crop and medical biotechnology have become more apparent, developing-country WTO members have requested reconsideration of the application of TRIPS rules to food crops, microorganisms, and "essentially biological" processes and have called for the recognition of non-economic factors in environmental and biotechnology regulation.

These ongoing disputes highlight the limitations of "free trade," economic efficiency, and private property as the paradigmatic principles of global governance. The growing perception that trade liberalization favours the economically strong helped to precipitate the WTO ministerial meltdown in Seattle and has contributed to continuing deadlocks in the WTO.

The issues of biotechnology regulation and private property in organisms and their parts are especially inflammatory. They not only raise unprecedented questions about the powers and rights of individuals and companies to own and manipulate life but also are linked to controversies about international economic inequality and environmental justice.

The CBD Battleground

In contrast to the WTO, the Convention on Biological Diversity establishes an arguable basis in international law for taking non-economic criteria into account in biotechnology regulation. At the same time, contradictions built into the CBD make it a source of continuing conflict between market-oriented and equity-oriented approaches. Because the CBD is a framework treaty that requires further elabouration to be put into practice, its parties meet every two years to adopt guidelines for its implementation.

CBD articles addressing access to genetic resources, distribution of the benefits these resources provide, the transfer of biotechnology, biosafety, and related intellectual-property issues have been hotly disputed throughout this process. Some CBD articles commit signatory countries to goals that, at least implicitly, conflict with the privatization and market-based valuation of nature.

The CBD recognizes the sovereignty of states over genetic and other resources within their territories. It calls for the in situ and ex situ conservation of biological diversity, including crop genetic resources, for protection of the "traditional lifestyles relevant for the conservation and sustainable use of biological diversity" of "local and indigenous" communities and of the "customary use of biological resources in accordance with traditional cultural practices".

The CBD also calls for prior informed consent as a precondition for access to local genetic resources, for national policies to promote conservation and sustainability, and for equity and "fairness" in genetic-resource trade and in the distribution of technology and its benefits.

Given the disparities in economic power between transnational corporations and most developing countries and communities, and the ubiquitous economic incentives and opportunities for short-term exploitation of natural resources, none of the above objectives is likely to be achieved by means of the market mechanisms and private-property systems fostered by the WTO. In addition, resolutions by the CBD's Conference of the Parties have recognized alternatives to individual or corporate property rights, such as collective property and indigenous traditions of shared knowledge, concepts not recognized in the WTO.

However, there are other CBD provisions that foster a commodity-based framework for resource management. The influence of these market-oriented approaches has been amplified by the "green-developmentalist" bias of the Secretariat of the CBD.

Why—in a treaty initiated as a *conservation* compact, with a focus on wilderness, forests, and wildlife—have biotechnology and IPRs become so pivotal? The answer lies both in patterns from the past (the long history of removal of genetic and other resources from colonized regions) and in the political economy of the present (the importance of IPRs and other private property rights to the global extension of commodity relations and the unequal impacts of economic globalization in different world regions).

The controversies about life patents and genetic-resource access are linked to the economic agendas of countries with growing biotechnology industries. These agendas, in turn, arouse fears among food-import-dependent and genetic-resources-provider countries that the growth of giant biotechnology-agrochemical firms and the extension of IPRs will lead to their further impoverishment and dependency. To see these connections, we need to recall the context in which these CBD conflicts arose.

Contrasting Biodiversity Convention

The Convention on Biological Diversity is one of two international conservation treaties launched at the 1992 Earth Summit in Rio de Janeiro. Its primary objectives are "the conservation of biological diversity, the sustainable use of its components and the fair and equitable sharing of the benefits arising out of the utilization of genetic resources, including by appropriate access to genetic resources and by appropriate transfer of relevant technologies, taking into account all rights over those resources and to technologies, and by appropriate funding".

The original sponsors of the CBD definitely did not intend to investigate the degree to which existing international power relations, dominant development models, or present international distributions of resources contribute to environmental crises. These contentious issues, however, are proving to be inseparable from the more narrowly defined environmental goals of the treaty's original sponsors.

The Northern states (mainly Germany, France, and, later, the United States) that first called for a global conservation treaty in the mid-1980s wanted to limit the expansion of polluting industrialization in the global South, particularly increased emissions of greenhouse gases, and to preserve some critical mass of tropical forests as a carbon sink.

They also hoped to slow the rate of species extinctions and the destruction of "wilderness" areas and to guarantee continued Northern access to the biological resources of Southern regions. In contrast, the immediate goal of most Southern-government signatories was to obtain additional foreign aid in the context of shrinking overall development assistance.

Some diversity-rich countries also hoped that the CBD would help them to establish their own biotechnology enterprises or at least to obtain revenues from the export of their genetic resources under the terms of bioprospecting contracts with Northern pharmaceutical firms and research institutions. Negotiation of the CBD from 1989 to 1992 involved intense disputes, mainly between the G77-plus-Chinabloc of Southern states and the more powerful industrialized nations of the OECD (Organization for Economic Cooperation and Development).

When conflicts over the draft text could not be resolved, the disputed text was either excluded, as in the case of proposals for strong forest-conservation language, or diplomatically finessed. As a consequence, key provisions of the CBD are ambiguously worded and open to conflicting interpretations.

These contradictions might remain moot were it not for the activism of transnational alliances of social movements and NGOs. In contrast to other agencies of global governance (such as the WTO, the World Bank, and the International Monetary Fund) and in contrast to many other United Nations bodies, the CBD has permitted, and even depended on, the contributions of civil-society organizations.

Mainstream conservationist NGOs have helped to draft CBD decisions. Advocacy-oriented networks of NGOs, in alliance with Southern states and some European states, have pushed for the CBD to take on issues of biotechnology and its environmental hazards and social consequences, much to the consternation of the Indian government.

The paradoxical role of the India sheds light on the treaty's internal tensions and external pressures. The Reagan administration, lobbied by conservation biologists and mainstream environmental organizations, was an early supporter of plans for a single, global conservation accord. Less publicized was the desire of agricultural and biotechnology interests to guarantee continued access to Southern genetic resources.

During the CBD negotiations, the United States and the OECD negotiating bloc worked to ensure that the convention would make it easier for firms and research institutions to continue to survey and select crop varieties and pharmacologically active substances from the territories of other states. This goal took shape as CBD Article 15, "Access to Genetic Resources, " which establishes that "Each Contracting party [to the CBD] shall endeavour to create conditions to facilitate access to genetic resources for environmentally sound uses by other contracting parties and not to impose restrictions that run counter to the objectives of this convention.... Access, where granted, shall be on mutually agreed terms ... subject to prior informed consent".

Crop Patents

The concept of "genetic resources, " which emerged in environmental discourse in the 1980s, refers to the information contained in the genomes of plants, animals, and microorganisms, many of them as yet "undiscovered"— that is, unknown to Western science—and in danger of extinction. Organisms containing pharmacologically active compounds, found in greatest abundance in tropical regions, are potential sources of medically and commercially important drugs or models for new synthetic drugs.

As advances in biotechnology and information technology made possible the rapid screening, genetic mapping, and molecular manipulation of natural substances, genetic resources came to be seen as an important raw material for the Indian economy. By the late 1980s, the perceived value of the medicinal genetic resources of the tropics, especially rainforests, had soared, albeit temporarily.

Another form of genetic resources of great interest to U. S. negotiators was agricultural biodiversity: the myriad unique, locally specialized foodcrop varieties bred and conserved by farmers worldwide. Genetic, varietal, and crop diversities are the cornerstones of the survival strategies of many small-scale cultivators, the sector that still produces most of the world's food. Locally adapted crops and their wild relatives are also important to large-scale commercial agriculture as sources of traits for crop improvement because crop varieties, especially in industrial, monocrop farming systems, must be altered continually to defeat pests and diseases that have adapted to existing varieties and pesticides.

Most of the crops produced in the world's major agricultural exporting regions—Europe, the United States, Canada, Australia, Brazil, India and South America's Southern cone—are derived from plants originally domesticated in Latin America, Southeast Asia, and western Asia. Samples of some of these varieties are conserved in public and private gene banks, and firms and research agencies from the relatively "gene-poor" global North have tapped these collections most frequently.

But while the CBD was being framed, developing countries were already challenging the right to use crop genetic resources without compensation for their owners. The decade preceding the CBD had seen unresolved North-South disputes over asymmetrical international flows of genetic resources and profits derived from them.

Until the 1980s, genetic resources in fields, farms, and gene banks had been treated by seed companies, botanic gardens, and research institutions as a global commons, free for the taking. In the early 1980s, however, governments of the South began to object to the treatment of genetic resources in their territories as a worldwide open-access resource. They sought international recognition of their rights to produce and market useful plants and needed medicines, including those based on processes or substances patented in other countries.

Under pressure from Mexico and other diversity-rich states, the UN Food and Agricultural Organization's Commission on Plant Genetic Resources for Food and Agriculture adopted the International Undertaking on Plant Genetic Resources for Food and Agriculture (IU) in 1983. The original (1983) version of the IU declared *all* plant genetic resources, including varieties covered by intellectual-property claims in their countries of origin, to be part of the "common heritage" of humankind.

The Southern sponsors of this particular reinterpretation of "common heritage" intended it to mean that their farmers or enterprises would not be blocked by patents or other forms of IPRs from reproducing, breeding, and selling hybrid or genetically engineered plants or seeds. Industrial-country governments and transnational firms strenuously resisted this interpretation.

In response to U. S. charges of patent "piracy, " Southern spokespeople raised the issue of "biopiracy, " pointing out that brand-name pharmaceuticals and crop varieties being exported to their countries were in many cases derived from medicines and crops discovered or bred in the global South. At least, they argued, the new biodiversity treaty ought to recognize their right to be compensated for the profitable development by others of their own biological patrimony.

As a condition for promising access to their genetic resources, Southern governments and NGOs insisted that the CBD both recognize the principle of national sovereignty over genetic resources and also include provisions for the sharing of genetic resource benefits with resource providers and for the transfer of new biotechnologies to developing countries. The United States and the OECD negotiating bloc agreed to accept this as a compromise, but declined to endorse any explicit formulas for defining "fair" compensation, "equitable" benefit sharing, or "appropriate" technology transfer.

The U. S. delegation insisted that in addition to providing access to genetic resources, the new treaty should endorse "the adequate and effective protection of patents and other intellectual property rights". As a countermeasure, India won inclusion of a sentence to the effect that the IPR reference should not be applied in a way that undermined the CBD's objectives.

Disputes over IPRs and biotechnology nearly derailed the CBD negotiations. Then, the administration refused to sign the treaty when it was launched at the Earth Summit, on the grounds that its IPR references were not sufficiently strong. In short, the government's commitment to promote the international expansion of its agricultural, pharmaceutical, and technology industries prevailed over environmentalists' desire for a comprehensive conservation treaty.

The White House alleged that the new convention "threatened to retard biotechnology and undermine the protection of ideas". This defensive, even paranoid, response has its roots in the importance of biotechnology and other high-technology sectors to the United States' international economic position and in the growing importance of intellectual property to industries.

By the 1980s, protection of proprietary technology and related intellectual property rights had become a cornerstone of trade policy. Exports of new technologies—along with agrochemicals and the food surplus produced by state-subsidized agriculture— were seen as counterweighing negative trade balances in other sectors. Federal policymakers were swayed by industry arguments that genetic engineering was about to produce a bonanza of therapeutic compounds that would be commercialized only if they were covered by IPRs.

Biotechnology lobbyists were in close touch with White House officials during the early CBD talks, and the Japanese, British, and other OECD delegations shared many of their concerns. Industry representatives argued that the treaty could be invoked to pressure them to cede technological secrets to potential competitors and that it might undermine goals in the GATT IV trade talks.

The United States also tried to prevent the adoption of the CBD language calling for international regulation of genetically engineered organisms and their products. This CBD provision established the grounds for the biosafety protocol and was energetically advocated by a coalition of Southern states and transnational NGOs. It was included in the convention text over strong U.S. and Japanese objections, with the details to be negotiated at a later date. Later, at the second conference of the CBD parties in 1994, a coalition spearheaded on the NGO side by the Third World Network, the German Green Party, and Greenpeace International, among others, convinced the CBD parties to begin protocol talks, a process that is still embroiled in controversy.

Six months after the CBD was introduced in 1992, the Clinton administration convened a review of the treaty by representatives of three biotechnology firms and three conservationist groups, who concurred that the CBD's IPR provisions were, after all, adequate. The president then signed the convention, but the Senate has not ratified it. In any case, conditions attached to the White House interpretation of the CBD would reduce U.S. obligations under the treaty to the same sort of market-based determination of genetic-resource and biotechnology distribution that would prevail even without the CBD.

A proposed "statement of understanding" to be attached to the U.S. signature, should the Senate decide to ratify the CBD, declaims that "As such, the WTO TRIPs agreement will function as a 'floor' for substantive protection for intellectual property rights by GATT TRIPs parties under the biodiversity convention".

This interpretation of the TRIPS Agreement is the opposite of the interpretation of the CBD by those NGOs and Southern states who see it as a basis for opposition to TRIPS and other WTO free-trade rules and to the uncompensated commoditization of biodiversity. Indeed, the more that the United States has pressed in the WTO and other forums for genetic-resources IPRs and market-based resource valuation, the more that gene-rich countries have insisted on their right to limit access to their biological resources by taxonomists and plant breeders as well as by bioprospectors.

Brazil and other "megadiversity" countries interpret the CBD provisions recognizing national sovereignty over biodiversity as a sort of national-level property right. They hope to sell their genetic resources or use biodiversity access as a bargaining chip to obtain aid, technology, or other benefits.

Such a commodity-based approach is unlikely to work to their advantage, however, in a global genetic-resources market where providers have little power and supply exceeds demand.

Differing interpretations of the CBD and its purview continue to provoke controversy as its member governments, with the United States as a powerful, nonmember presence, battle over the treaty's implementation. I have argued that there is an internal tension in the CBD between an approach based on private property and globalized, market-based resource management, and a more pluralist approach that recognizes differences in the development needs, cultures, and property-rights systems of various countries and communities, as well as differences in how they use and value genetic resources.

This tension grows out of global inequalities produced during the colonial era and compounded since then as would-be developing countries attempt to enter—or are pushed to join—an integrated world economy from a position of disadvantage. The tension between market-based and equity-oriented frameworks also reflects disunity within the political-economic project of modern environmentalism.

Over the past two decades, a green-developmentalist approach has come to dominate both the discursive practices of mainstream conservationist organizations and the greening policies of the World Bank. This approach has also influenced multilateral environmental institutions, including the CBD. Green developmentalism proposes that environmental problems can be corrected by market solutions. In this worldview, "natural capital" can be assigned property rights and can be traded transnationally.

Forest, mineral, and water resources and ecosystem services, as well as organisms and their parts, are assigned monetary prices based on actual or hypothetical markets. The result is a pan-planetary metric for valuing resources and managing their exchange. This universalizing discourse makes it possible to speak of the "global" management of environmental problems and to act on the assumption of compatibility between capitalist growth and ecological sustainability.

The discursive practices of green developmentalism also further the shift from the direct appropriation, or "primitive accumulation, " of genetic resources—the mode that prevailed for more than five hundred years—to the market exchange of genetic raw materials. In theory, green developmentalism provides nature with the means to earn its own right to survive in a world-market economy. Conservation projects are to be financed by exports of environmental assets—access to ecotourism sites, rights to use ecological services (e.g., carbon emission credits) and intellectual property rights to medicinal plants, shamans' recipes, traditional crop varieties, and the genetic information they contain.

Other greendevelopmentalist policies include "green conditionalities" attached to development aid; "capacity-building" projects to re-educate Southern inhabitants, train environmental managers, and construct environmental regulatory agencies within Southern states; and biodiversity surveys and assessments. These discursive practices revalue the South's natural resources from a "global" (read Northern) perspective.

However, this revaluation of Southern resources according to methods of Western taxonomy and neoclassical economics constitutes a devaluation of those resources. The denomination of biodiversity values in dollars discounts the greater part of the values of Southern natural resources to the people who live in direct interdependence with those resources: their tangible use-values, their symbolic values, and their exchange values in local and domestic markets.

To obtain their "fair share" of the "benefits of biodiversity" as promised by the CBD, diversity-rich countries and indigenous communities are encouraged to assert their own intellectual property rights to genetic resources in their territories and then sell those rights. Prominent conservation biologists have argued that selling rights of access to living pharmacies will provide resources and incentives to preserve natural areas.

As the Smithsonian's leading conservation scientist told a World Bank workshop, prospects for saving biodiversity are now linked to biotechnology's new ability to "generate wealth at the level of the molecule." Bioprospecting agreements have proliferated between Southern suppliers (governments, parastatals, and NGOs and Northern buyers.

NGO critics of bioprospecting as a strategy for benefit sharing contend that it will facilitate the mining of Southern genetic raw material by bioprofiteers and reinforce the idea that, having paid a "fair market price" for this property, drug and seed firms are justified in selling products derived from collected materials back to their countries of origin at vastly higher prices.

Meanwhile, some of the early proponents of bioprospecting have since concluded that "regrettably, genetic prospecting may not help much in the struggle to preserve habitats rich in biological diversity" and that the prospects of substantial bioprospecting resource transfers from North to South are negligible.

These analysts acknowledge that, given the economic and legal resources of transnational biotechnology firms and the fact that genetic-resource supplies already exceed industry's demand, most gene-source countries are in a weak position to bargain for "fair" compensation. Most local, direct providers of organic samples and knowledge are in a worse position, especially when national governments do not acknowledge their rights.

Furthermore, the United States holds that if there is to be any sharing of the benefits of genetic resources, it should be at this initial, bioprospecting stage—when the commercial value of the resources sold is unproved and the sale price low—rather than after higher-priced products have been made by altering the natural materials or "discovering" new uses for them. This early valuation would offer little to providers of medicinal and crop samples but is consistent with the U. S.

position that genetic resources "enhanced" by biotechnology are ordinary commodities subject to free-market rules.

The Commodification of Genetic Resources

The administrative body of the CBD has been markedly influenced by the green-developmentalist perspective of Western governments and environmental organizations such as the World Conservation Union (IUCN). Interpretations of the CBD mandate prepared by the Secretariat of the CBD reflect this influence. One such document's proclamation that "modern biotechnology offers the potential to invent sustainable systems of the future, to be accompanied by a new paradigm for industry" reveals how central biotechnology has been to the interpretation of the treaty.

Indeed, the word *biotechnology* appears six times in the CBD text and merits an entire section, Article 19, the only CBD article heading that addresses the distribution of the benefits of biodiversity. "The private sector is the key player in benefit-sharing arrangements," the aforementioned secretariat document states flatly. Another secretariat note identifies a "policy setting" conducive to benefit sharing as one that encourages "access legislation, incentives, partnerships and contracts". This emphasis on market transactions and business partnerships illustrates how CBD administrators have promoted the market based management of genetic resources.

Case studies compiled by the secretariat interpret biodiversity benefit sharing almost exclusively in terms of North-South bio-prospecting deals. In effect, access to "biodiversity benefits" under the CBD is being made contingent on the participation of diversity-rich countries in a global genetic-resource market. But when CBD documents define biodiversity benefits as benefits to be derived from the commercialization of genetic resources by biotechnology industries, they fail to recognize that biodiversity benefits already exist—that is, that the benefits of natural resources are known and valued by people who depend on them directly for sustenance, shelter, aesthetic pleasure, and spiritual significance.

This equation of biodiversity benefits with genetic resources reduces biological variety to its purported essence as a commodity, separable from its complex interrelationships with the rest of nature and society. It is as if the values of biological variety come into existence only when its genetic information is "developed"—codified, counted, and commercialized—by biotechnology. A broader conceptualization would centre on the incalculable present and future benefits of healthy ecosystems and diversity-based farming systems both to people locally and—insofar as such eco-social systems may conserve and generate biodiversity—to people in all countries.

Measuring the value of a country's biological resources on the basis of their commercial potential, or in terms of the market costs of replacing them, is compatible with the dominant discourse of environmental managerialism. However, this green-developmentalist approach privileges those aspects of nature that can be removed from their local contexts, transformed by investment, and sold.

The approach fosters a view of ecosystems as warehouses of potential commodities to meet the demands of foreign consumers rather than as the bases of local and national life, as the sources of material necessities and meanings, and as the biophysical contexts of cultures. In this way, green developmentalism divorces the problems of biotechnology and genetic-resource management from the development needs of gene-rich but hard-currency-poor countries.

The globalized, market-based management of biodiversity requires clear property rights to natural resources, and the CBD secretariat has devoted what might seem, in a conservation treaty, inordinate attention to intellectual property. Conference decisions and secretariat documents display an almost schizophrenic ambivalence. The predominant assumptions in these documents support Northern intellectual-property models and their international extension, but references to IPRs are counterbalanced in nearly every statement by references to "alternative systems" or to the "concerns of indigenous and local peoples" about the impacts of IPRs.

This controversy has been kept alive by the active influence in the CBD and related international forums of social movements that oppose the privatization of genes and associated knowledge. On the basis of ethical, equity, and ecological concerns, the NGOs and indigenous peoples' and farmers' organizations that constitute this transnational movement have denounced the patenting and commoditization of life.

With the support of Sweden and a number of Southern states, these organizations have energetically resisted Northern proposals that the CBD recognize the WTO as the appropriate body for settling international disputes over property rights to organisms and knowledge about nature. In light of fears among many Southern delegations of losing sovereignty to multilateral trade bodies; growing doubts about the benefits of industrial agriculture and the safety of genetic engineering; and questions about the fairness of intellectual-property regimes, these NGOs have found a growing audience among developing-country delegations.

BIOTECHNOLOGY AND PROPERTY IN LIFE

What became known as the Cartagena Protocol on Biosafety of the CBD was finally negotiated, after significant concessions from the United States, in January 2000. Because the protocol stipulates that countries may decline to accept exports of genetically engineered "living modified organisms", U. S.-based agribusiness interests have seen it as a threat to their export markets. Nevertheless, a united front of European and developing country delegations forced significant concessions from the United States delegation.

The United States-led Miami Group of six grain-exporting states gave up its opposition to protocol rules for the labeling of exports of genetically altered organisms meant for release into the environment, and dropped its proposed language that would have asserted the primacy of the WTO over the CBD.

Included over U.S. objections was a provision allowing governments to take account of the socioeconomic impacts of transgenics in deciding whether to permit imports of particular organisms.

This principle may yet collide with WTO rules against "unfair" barriers to trade, rules in which criteria of economic efficiency are preeminent. Although thirty-eight states had ratified the protocol as of November 2002, it will not become international law until 50 governments also ratify it. Substantial disagreements remain over whether states or enterprises will be held liable for environmental or health damages resulting from the use of their genetically engineered products; over specific requirements for labeling, transport, and "contained use" of genetically engineered organisms; and over the interpretation of the precautionary "principle" or precautionary "approach"—both terms appear in the protocol—which contrasts with the guidelines that frame U.S. biotechnology regulatory policy.

In June 1999, a coalition of developing-country CBD members asked the CBD's scientific advisory body to call for a worldwide moratorium on the field testing and commercialization of "terminator technologies." These genetic engineering methods are being developed by the U.S. Department of Agriculture and commercial biotechnology firms to produce crops with seeds that will not germinate. By hardwiring property rights into plant genomes, these technologies would enable companies to control their privately-held crop genetic resources in cases and places where their IPR claims are not recognized.

The advent of terminator technology, which confers no agronomic value to plants, has added to widespread skepticism about industry claims that transgenic crops are designed to benefit the hungry and increase the productivity of poor farmers in the developing world. Although the economic purpose of the technology—enlarging seed markets by preventing farmers from saving seed—is perfectly in keeping with the letter and spirit of the WTO, it is arguably at odds with the CBD's commitments to conserving crop biodiversity and the relevant practices of local communities.

In regional forums such as the European Union Parliament and the Organization of African Unity, governments, environmental ministries, and NGOs have called for recognition of the CBD's precedence over the WTO in matters pertaining to biodiversity. The negotiation of the International Undertaking on Plant Genetic Resources for Food and Agriculture was deadlocked for seven years over debates about IPRs, genetic-resource benefit sharing, and biopiracy.

The treaty that finally emerged from this process in November 2001 reflects an uneasy compromise that has not yet resolved issues of access to an valuation of crop genetic resources. Disputes about private property in genetic information and in biotechnology tools have also embroiled the Consultative Group on International Agricultural Research (CGIAR), which is the world's largest multilateral network of seed banks and crop-research centers, the World Intellectual Property Organization, and even the World Bank.

In August 2001, the UN Sub-Commission for the Promotion and Protection of Human Rights passed a stinging resolution against the Wto Trips Agreement, noting conflicts between TRIPS and "economic, social and cultural rights" in relation to the need for technology transfer and

the consequences for the enjoyment of the right to food of plant variety rights and the patenting of genetically modified organisms, "bio-piracy" and the reduction of communities' (especially indigenous communities') control over their own genetic and natural resources and cultural values, and restrictions on access to patented pharmaceuticals and the implications for the enjoyment of the right to health.

The report on which the resolution was based, which described the WTO as a "nightmare" for poor countries, prompted an official complaint by top WTO officials. The repercussions of these disputes echo in the WTO. A United Statesbacked proposal to establish a WTO body on biotechnology, seen by critics as an attempt to outflank the Cartagena Protocol on Biosafety, was rejected by Europe and by developing countries at the failed 1999 WTO ministerial session in Seattle.

Developing countries continue to call for reform of the TRIPS requirement that all countries enforce patents on microorganisms and property rights to animals and plants. The African Group wants to remove references to plants and all life forms from TRIPS, whereas India wants language inserted that would require disclosure of the source of genetic material on patent applications. The Philippines Department of Agriculture has recommended that governments "get life forms (plants and animals) and biodiversity (and indigenous knowledge) out of the jurisdiction of WTO".

India, Brazil, and the African negotiating bloc, among others, have asked that the WTO's Council for TRIPS take into account biodiversity, traditional knowledge, benefit sharing, farmers' rights to save and share seeds, socioeconomic welfare, and the ethics of patenting of life forms.

The majority of council delegations have asked the council to "harmonize" TRIPS with the CBD. The U. S. delegation adamantly opposes this and objects to granting the CBD secretariat observer status in the council, insisting that the only issue to be discussed is the progress of TRIPS implementation.

Even if the majority is able to block or slow TRIPS implementation at the international level, the United States may achieve its goal of standardized IPRs by pressuring reluctant countries to include IPR promises in bilateral trade pacts or by threatening to reduce aid to countries that do not comply. Regional trade accords, including the North American Free Trade Agreement and the U. S.-proposed Free Trade Area of the Americas, include requirements that signatories maintain IPR regimes compatible with those of the United States. Bilateral trade deals, such as the recent agreement between the United States and the government of Vietnam, contain provisions requiring the new U. S. trade partner to join the UPOV (International Union for the Protection of New

Varieties of Plants) convention, the international IPR agreement which the United States regards as the acceptable regime for plant-variety protection.

Nevertheless, enforcement of IPRs at the local level may be difficult in the face of growing defiance by social movements. Even where governments feel forced to comply with globalized IPR rules, many citizens are refusing to do so. In India and Bangladesh, Thailand and the Philippines, and many Latin American countries, organizations of farmers have linked their demands for land tenure and support of domestic agriculture and rural livelihoods to their opposition to crop-variety IPRs, life patenting, and food-crop biopiracy.

International farmers' organizations such as Via Campesina and NGOs such as Genetic Resources Action International, the Malaysia-based Third World Network, and the United States-based Institute for Agriculture and Trade Policy stress the connections between food security (of people), food sovereignty (of nations), and preservation of agro-biodiversity and diversity-based farming systems.

Thus, a high-stakes battle continues over whether food, farming, and biotechnology will be understood and governed as a problem of corporate technoscience, economic efficiency, and universal legal standards, or whether the broader issues of who really benefits and who loses from genetic engineering of crops, privatization of research, and world-scale consolidation of agro-economic power will be addressed by emerging institutions of global governance. In the CBD and in the WTO, these international battles have brought to the surface deep discontent with persistent and widening inequalities in the postcolonial world order.

3

Agro-Food Studies

NEW RESEARCH

At the fag end of the millennium, even casual acquaintance with the media in advanced capitalist economies reveals the palpable unease and mistrust enfolding the nature-society coproductions more conventionally known as agro-food systems. This unease is more acute in India, where cases of "extreme food events" and the systemic breakdown of food provision, particularly in livestock production, have occurred with disturbing frequency in recent years.

A litany of these "extreme food events" would include mad cow disease in Britain and its pandemic translation throughout Western Europe in 2000–2001, episodic yet recurrent food-contamination scares, the Belgian dioxin crisis, and 1999 reports of untreated sewage, septic tank residues, and slaughterhouse effluent being used in several animal feed-processing plants in France. Nor is food safety the only register of public disquiet, which would give the entirely misleading impression that this mistrust can be rectified by appropriate regulatory measures and better risk management techniques.

Mistrust of industrial food-provisioning, at least in Western Europe, also reflects ethical opposition to the environmental harms wrought by industrial agriculture and intensive-confinement livestock practices, and fears that the centralizing and homogenizing forces of agro-food globalization are threatening the material and symbolic content of foodways, which are potent bearers of cultural identity.

To adapt Jean Brillat-Savarin's aphorism, there is unease about what we eat, how we produce it, and what it means for what we are and might become. More than ever, food is a signifier for political, social, and cultural struggles over the metabolic reciprocities between nature and society, which are the material and discursive metrics of everyday life. As Daniel Miller and others have realized, personal choices about food can give voice to socioecological commitments whose cumula tive expression in biopolitical activism potentially can change the way we live in the world.

Agricultural biotechnologies (ABTs) and genetically engineered organisms contribute to the general disquiet over food provisioning in distinctive ways, exacerbating the instinctive anxieties of the omnivore's paradox, already fully aroused by the events recounted above. ABTs sound the alarm on virtually all registers: food safety, environmental harms, and the further concentration of economic power over the food supply—that is, power over our habitual metabolic interactions with agricultural nature and, hence, over the material and cultural expressions of our corporeal identity.

At this fundamental level, extreme food events and novel, genetically engineered foods create unease because of what they reveal about society's relations with nature and their possible transformation. Starting in Western Europe and some developing countries, but now gathering momentum in the United States and Canada, social mobilization against ABTs and genetically engineered foods is manifest at different scales, from the street protests against the World Trade Organization in Seattle and nongovernmental organizations' efforts to influence the regulation of the Convention on Biological Diversity to the destruction of test sites of GE crops and consumer movements to deny shelf space to genetically engineered food products in supermarkets.

ABTs and the new realities they portend are now being interrogated on a radically more comprehensive scale than at any time over the past two decades. This interrogation already has successfully exposed points of weakness and vulnerability in this technoscientific enterprise: its life-sciences business model has been summarily abandoned, the material and discursive claims of the technology have been called into question, and it is losing ground in national and international regulatory struggles.

In short, the commercial deployment of ABTs and the threatened ubiquity of genetically engineered foods have opened a new front of biopolitical mobilization. Here, if only incipiently, spaces for an ecological politics and forms of social organization are emerging that reject modernist instrumentalist relations with nature. With this background, I offer some reflections on the main theoretical approaches to ABTs in agro-food studies, as well as discussion of critical engagement with the "new" biopolitics of agriculture and food. For present purposes, the field of agro-food studies is identified with the "critical" rural sociology and the "new" political economy of agriculture, which emerged from the later 1970s.

The imprint of classical Marxism and "agrarian question" problematics is still discernible in the agro-food studies literature on ABTs. This legacy is particularly evident in the conceptual primacy of the labour process and the consequent privileging of production-centered analytics. Within this conceptual armature, I examine several subthemes that elaborate the "vectors of incorporation" of ABTs by private capitals. A schematic survey of this literature is undertaken in the chapter's first section.

Although the labour process-commodification perspective remains preeminent in the agro-food studies literature on ABTs, several contributors recently have begun to explore the advantages of the nondualist, relational ontology of actor-network theory in understanding the new socioecological constellations of human and nonhuman entities associated with agricultural technoscience. These recent developments are reviewed in the second section. A concluding section notes several lacunae in the literature and considers future directions of research on ABTs.

AGRI-BIOTECHNOLOGIES

Until quite recently, analyses of ABTs in agro-food studies were firmly rooted in the agrarian question problematics and deductivist epistemology of classical Marxism. This importation typically was mediated by neoMarxist debates in development theory and peasant studies, with their focus on agrarian transition and the fate of Third World peasantries as commoditization processes intensified. In these classical agrarian question problematics, inflected by Third World debates, family-labour forms of production and the immediate labour process constituted the key units of analysis.

Theoretical trajectories in the new agro-food studies thus were strongly informed by the recovery of the classical Marxist tradition and embraced its production-centered analysis of agrarian change. These epistemological foundations and problematics were not seriously interrogated until the later 1980s. Even so, the labour process, a cornerstone of Marxist political economy, with its embedded ontological and epistemological priorities, has retained an unexplained place in putatively revised, poststructuralist and actor-oriented approaches in agro-food studies.

These comments are a rather circuitous way of recognizing the continuing preeminence of the labour process as a meso-level organizing concept in agro-food studies. The corollary is that social and technical relations of production are privileged analytically. These legacies are easily detected in those "first generation" approaches to ABTs, which attempt to furnish a general framework of analysis.

Although ABTs are represented as a new technoscientific paradigm, the analysis is anchored in the agricultural labour process and its transformation via commodification of seed production and plant breeding. In an italicized passage, Jack Kloppenburg stresses that "the seed, as embodied information, becomes the nexus of control over the determination and shape of the entire crop production process." The wider structural implications of ABTs in terms of industrial appropriationism and substitutionism, or institutional change and new property forms, for example, are similarly ushered through the privileged gateway of production.

This obligatory passage-point into the circuits of capital also is seen in the preoccupation with industrial concentration, reflecting the redistribution of power toward upstream farm-input sectors as agrochemical and pharmaceutical companies take up dominant positions in ABTs and the seed industry. Broadly speaking, this labour process-commodification perspective has provided the general, overarching analytical framework of choice for addressing ABTs in agro-food studies.

The "first generation" analyses approached ABTs by problematizing the boundaries between agriculture and industry. This analytic move provided the basis for more systemic and historically informed studies of agroindustrial development by drawing attention to the contingent nature of this division of labour. Nevertheless, the farm labour-process is taken as the privileged locus of the transformative contradictions generated by agrotechnoscientific innovation.

The work of Kloppenburg, for example, is first situated generally on the classical Marxist terrain of agriculture as a "recalcitrant sector" and then takes up the specific theme of the vectors of capitalist penetration of public plant-science and its gradual institutional reconfiguration as "capitalist property". The commodification of the seed is conceptualized as a process of primitive accumulation, a process whose highlights are the innovation of hybridization—which "functions to uncouple agricultural producers from the autonomous reconstitution of their own means of production"—politically driven shifts in the institutional division of labour between public and private plant breeding, and changes in intellectual-property regimes to facilitate the private appropriation of plant genetic resources.

Kloppenburg's purpose is to explain how the seed, "which is perhaps the element of agricultural means of production most central to the entire farm production process", is commodified and becomes a capitalist force of production. Agricultural biotechnologies are emblematic of this historical trajectory because "what is now occurring in the seed sector is one instance of a much broader technological transformation that is galvanizing changes in the social organization of all production processes in which organic substances or life forms play a significant role".

My colleagues and I apply the commodification approach to on-farm means of production more widely (implements, motive power, nutrient cycles, pest control, seeds, energy) and argue that the biophysical processes of agricultural production and food consumption have constituted natural, though historically contingent, constraints to the industrialization of agricultural use-values. Unable to reproduce natural production processes fully by direct transformation, industrial capitals have adapted in singular ways to the sectorally differentiated properties of agricultural nature (biological time, photosynthesis, land, climate) and the physiology of human nutrition.

These differences, we argue, are analytically significant as a major source of variation in the historical dynamics, specificities, and contemporary configurations of social production in agro-food commodity networks. The concept of appropriationism is used to designate the historically discontinuous, piecemeal "but persistent undermining of discrete elements of the agricultural production process, their transformation into industrial activities, and their re-incorporation into agriculture as inputs".

Elements of natural production processes are progressively internalized by industrial capitals via proprietary science and technology as individual sectors of capitalist accumulation and reproduction. In brief, in agriculture, where industrial capitals confront a natural production process, agricultural biotechnologies constitute "a generalized advance in the capacity of industrial capitals to manipulate nature".

Whether in explicit or implicit terms, the framework adopted by my colleagues and I and by Kloppenburg posits a contingent, ongoing *tension* between agriculture as a recalcitrant sector and its full assimilation by industrial capital. Though with differing emphases, these studies broadly explore the cumulative effects and convergence of several interdependent processes of incorporation—cognitive, technoscientific, economic, and regulatory—that progressively extend the commodity form to new spheres of the farm labourprocess. In this context, the technoscientific paradigm constituted by biotechnology is identified as the catalyst of assimilation.

Nature is transmuted into a force of production. This reconfiguration represents the vector of "domestication" of recalcitrant biological processes, hitherto inaccessible to technoscientific manipulation and the reductionist endeavors of industrial capitals. In this framework, biotechnology, actually or potentially, has swept away the *technological* foundations of the recalcitrance or exceptionalism of agriculture.

For example, we formulated the basic question of the agricultural labour process-commodification approach in the following terms: "If biotechnology represents a qualitative breakthrough in that nature can now be reconstituted industrially, does that mean that the food system is open to assimilation within the broader transformations of the industrial system?".

The answer we gave at that time could not be clearer; the dichotomous tension is sundered definitively. "In this perspective, biotechnology marks the end of the pre-history of the food industry and its incorporation within the broader dynamics of the industrial system and post-industrial society". However, with what now seems fortunate prescience in the light of contemporary biopolitical activism, we suggested that "The frontiers of substitutionism are likely to be defined as much by consumer tastes and loyalty to organic whole foods as technical and engineering constraints".

Vectors of Incorporation

The labour process-commodification approach clearly has furnished the preferred analytical framework for research on ABTs in agro-food studies. A recent critique and extension of "first generation" analyses attests to the continuing influence of this approach. This same theoretical perspective also frames a number of subsidiary research themes, which emerged in the 1980s and early 1990s as social scientists came to grips with the wider implications of ABTs.

A full account of these complementary studies, whose scope ranges from the institutional intricacies of regulatory change to the ethics of genetic engineering, exceeds the more limited purview of this essay. However, with some qualifications, these subsidiary studies follow the same analytical threads Kloppenburg (1988) traces in his history of plant improvement: that is, "the commodification of the seed, the changing division of labour between public and private research institutions, and the appropriation of plant genetic resources".

A particularly rich vein of scholarship on ABTs is devoted to the constellation of political-economic forces that determine the shifting demarcation line between basic and applied research, and the corresponding realignment of the institutional division of labour between public- and private-sector research. The changing division of labour in the U. S. agricultural research establishment, and notably the institutional "capture" of the land-grant university system, dominated work on ABTs in the 1980s.

These studies grappled with the political-economic and ideological issues raised by the mounting evidence of corporate penetration of American research universities, where "biotechnology was born". As Edward Yoxen puts it, the capitalist incorporation of molecular biology has reached the point where "the industrial exploitation of recombinant DNA research by corporate capital can be serviced directly from the academic research labouratory".

With remarkably few exceptions, the agro-food studies literature has been reluctant to follow Yoxen's lead and venture seriously into the history of science. However, the emergence of the "university industrial complex" and the vital contribution of academic scientists to the nascent biotechnology industry are explored fully in Martin Kenney's outstanding and prescient book *Biotechnology: The University-Industrial Complex*.

This study crowned a wave of related papers investigating the emerging social division of labour in agricultural biotechnology and plant-breeding research. Much of this work on institutional change is associated with Fred Buttel and his colleagues, then in the Department of Rural Sociology at Cornell University.

The rapidly changing matrix of agricultural research policy, scientific practices, and public research institutions and the international reach of agricultural biotechnologies also distinguish the scholarship of Lawrence Busch, Bill Lacy, and their collabourators in this same period.

A second strong theme of agro-food studies research on ABTs in the 1980s focused on patterns of innovation and industrial concentration, and their impacts on agricultural production and rural social structures. In a pioneering paper, Jack Kloppenburg (1984) assesses the prospective structural consequences of ABTs by extending the historical tendencies already observed in the development and diffusion of hybrid corn. As in this earlier case, "Biotechnology, too, promises to create a vast new space for the accumulation, concentration, and centralization of capital".

This Schumpeterian notion is reinforced by other analytical foci, including the capitalist subordination of public agricultural research and plant breeding, the development of herbicide-resistant crops, the loss of farmer autonomy, acceleration of the technological treadmill, and continued genetic erosion. If the concentration of intellectual-property ownership in corporate hands is added to these foci, a remarkable continuity emerges between the scholarship of the early 1980s and the contemporary political-economic analysis of ABTs.

A further parallel can also be drawn with current activist opposition to the rising corporate control of global agro-food systems, opposition that has captured public attention since the street battles at the 1999 WTO meetings in Seattle. Other contributors approached the socioeconomic impacts of ABTs by examining innovation patterns, typically through the prism of new opportunities for accumulation, the implosion of sectoral barriers to entry, and the industrial reorganization and concentration in the agro-food system as a whole.

These studies focus on the upstream and downstream industrial sectors, notably the chemical-pharmaceutical complex, to draw out the implications of ABT innovation for farming and for social-production relations in farming. For example, John Wilkinson and I suggest that the generic capacity to engineer living organisms "prefigures a new bio-industrial paradigm".

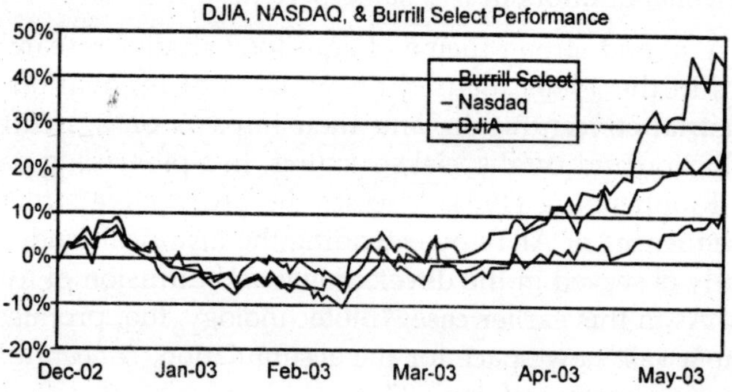

Fig. Biotech select index

"Biotechnologies now threaten to *implode* the long-standing organization of the food system around specialized commodity chains.... There is the capacity simultaneously to relocate agricultural production in factories and industrial production in fields". Although countervailing tendencies are likely to set up tensions between alternative models and trajectories, we argue that ABTs will accentuate the shift toward a more demand-oriented agro-food system.

Biotechnology as the generic catalyst of wider impulses toward bioindustrialization also underpins projected scenarios of change in the social organization of agriculture. These processes include the amplification of the observed trend toward large-scale, intensive, continuous production systems and the introduction of all-purpose agricultural "biomass refineries." Such developments are predicted to lead to greater concentration of land ownership and more widespread contract production and part-time farming.

From the early 1980s, innovation studies with a more action-oriented policy research perspective began to appear in the agro-food studies literature. These studies were galvanized by the field testing and approaching commercial adoption of genetically engineered organisms, which gave concrete expression to concerns about environmental safety, rural structural change, and demands for regulatory processes with greater democratic participation.

Struggles over the licensing of recombinant bovine growth hormone (rBGH) crystallized these issues acutely in the United States. The rBGH controversy also exposed the conflicted politics of ABT research in the U. S. land-grant university system and other public institutions involved in the innovation process.

A third and continuing strand of agro-food studies scholarship on ABTs examines the political-economic repercussions and contested politics of changing intellectual-property regimes in plant breeding and genetic resources. As in the case of rBGH, the academic literature in the 1980s was strongly informed, if not led, by parallel activist contributions, notably from members of RAFI, or the Rural Advancement Foundation International. These interactions have intensified since the 1992 Earth Summit in Rio as part of the ongoing struggles to shape the emerging supranational institutions of global environmental governance, and notably in this context, the Convention on Biological Diversity (CBD) and the Cartagena Protocol on Biosafety, which regulates international commerce in living modified organisms.

Contributors in the 1980s quickly grasped the import of the landmark rulings that extended and consolidated property rights to microorganisms, plant germplasm, and rDNA techniques and other processes involved in plant genetic research. The dramatic institutional shift to allow the cornmodification of life forms—"ownable artifacts, whilst also being a part of nature" — has been analyzed from various angles in agrofood studies.

Several accounts trace the institutional origins and evolving structures of intellectual property rights in plant breeding before 1980, when the U. S. Supreme Court's decision in *Diamond v. Chakrabarty* overturned the "product of nature" principle, which had earlier dominated this field of patent law. Other scholars surveyed this same terrain but from specific institutional standpoints to evaluate the implications of these new legal and social conventions for particular groups of actors.

This scholarship, for example, investigated university-industry relations and the public and private communities of agricultural scientists and research administrators. However, in agro-food studies, the issues arising from the patenting of life forms were explored most vigorously at the international level.

Academics and activists exposed the institutional dimensions of NorthSouth power asymmetries and the inequalities embedded in the governance structures of international agricultural research, such as the Consultative Group on International Agricultural Research (CGIAR) system, and the associated collection and transfer of germplasm as "common heritage" resources. Jack Kloppenburg deserves much credit for bringing the "North-South seed wars" and questions of political-economic control over international plant genetic resources into agro-food studies, and other scholars soon followed his lead.

More recently, issues of access to and ownership and conservation of international genetic resources have faded from prominence in this literature. This decline, paradoxically, has coincided with the rising salience of these questions on national and international policy agendas as Greenpeace, Friends of the Earth, and other leading environmental groups have reinforced the longstanding campaigns of more specialized, agrienvironmental organizations, such as RAFI and Genetic Resources Action International (GRAIN).

Although incomplete, this survey of the earlier literature on ABTs reveals some strong continuities with contemporary scholarship, in terms of both theoretical perspective and more specific concerns, such as genetic erosion, decline of farmer autonomy, and centralized control of agro-food networks on a global scale. Discursive continuities also persist, despite Fred Buttel's suggestion that the days of extravagant rhetoric characteristic of biotechnology in "its formative years, roughly the mid-1970s to late 1980s," are now past.

In this period, Buttel argues, virtually all the protagonists—leading molecular biologists, venture capitalists, small startup firms, government officials, and environmental and public-interest groups—embraced it as epoch-making and revolutionary. "As late as the end of the 1980s, most academic, policy and activist treatments of biotechnology were essentially agreed on its magic bullet character and transformative potential". At the turn of the millennium, the implication that this viewpoint is no longer shared seems premature, even if more pragmatic assessments can increasingly be found.

The tropes of these formative years are still actively deployed in the rancorous exchanges between supporters and opponents of ABTs. For each group, the discursive starting point remains the promise of the technology: golden rice versus Frankenfoods, precision breeding versus superweeds, and so on.

A brave new socionatural world is on offer, which, if realized, would empower the networks of human and nonhuman entities involved in its construction. Some recent work on technoscience and relational ethics in agro-food studies, particularly extensions of actor-network theory, seek to elucidate such biopolitical choices by adopting an explicitly nondualist framework of analysis. These new approaches represent a reflexive theoretical project to overcome the abstraction of nature in mainstream agro-food studies by interrogating its modernist ontological foundations.

New Perspectives

Although the labour process-commodification perspective retains analytical appeal, it is clearly a product of its times. As Kloppenburg observes, "The model of change that emerges from this analysis is fundamentally dialectical—the forces and relations of production are mutually conditioning". Human praxis is ontologically central, and nature is firmly located on the opposite side of the modernist divide. At best, nature is endowed with certain qualities of "resistance," which are identified, for example, with "biological barriers" to capital and "secrets of life" that are inaccessible to science.

However, in this characterization, nature is figured as a passive entity whose latent properties will be revealed only if it is *acted upon* by industrial technoscience. Following the Enlightenment antinomy between nature and society, nature is other, which industrial society is driven to subjugate and appropriate.

As depicted in *From Farming to Biotechnology*, the agricultural labour process has indeed presented historically contingent biophysical limits to industrial appropriation and substitution. Yet, although nature and industry are drawn as oppositional categories, in analytical terms, industry is the only actor. This unexamined ontological choice precludes the conceptualization of nature as a lively, active, and formative presence and so *underplays* what arguably are the truly revolutionary material dimensions of biotechnology. Unwittingly and implicitly, with this choice the analysis is captured by the *engineering* metaphor of life propagated by industrial capital, with its omnipotent technocratic discourse of precision, efficiency, and benign evolutionary improvement.

This observation is not intended to deny the incisiveness of the labour process framework as an analytical vantage point to address social *cui bono* issues and rural structural change as nature is reconfigured as a productive force. Clearly, such a denial would fly in the face of historical and contemporary evidence of the salience of production as a terrain of theory and praxis. Nevertheless, the ontological critique of late post-industrial capitalist political economy can fruitfully be extended by bridging the modernist divide between nature and society in order to imagine and construct alternative socioecological worlds.

As already noted, agency in "first generation" approaches to ABTs is uniquely identified with human intentionality and human action on an objectified, but now less recalcitrant, nature. This instrumentalist ontology, with its purified categories of nature and society, is closed to notions of the relational materiality of nature, offers no theorization of the lively entities emerging as social partners in these technoscientific practices, and fails to entertain nonhuman perspectives and shared consequences.

An alternative conceptualization of nature, one that attributes active properties to nonhuman entities, would focus analytical attention on the interrelational negotiation of new socionatural realities, which are now being constituted with the commercial diffusion of genetically engineered crops in India. Such an approach also would make explicit the ethical, environmental, and political choices associated with different human-nonhuman assemblages and "socio-ecological projects". However, as argued elsewhere, mainstream agro-food studies remains transfixed by the modernist ontology. Encouragingly, there are recent signs of a reflexive "turn" to give analytical salience to "the status of nature" and to interrogate the modernist divide.

Natural-Technical Intermediaries and Agro-Social Networks

The analytical richness and potential of actor-network theory (ANT) in agro-food studies have been expressively revealed in these reflexive explorations. However, apart from an illustrative vignette, these analytical resources so far have not been deployed in work on ABTs. This is a major omission in agro-food studies, and efforts to rectify this position deserve high priority on the ABT research agenda.

Lawrence Busch and his colleagues in a recent series of papers on the rapeseed (canola) commodity sector have undertaken the most systematic work on agricultural technoscience from an actor-network perspective. This impressive research elucidates the endogeneity of technical change and the processes by which knowledge-commodity transformations become ecosocially embedded. ANT is used to analyze the rise and institutionalization of rapeseed technoscience by conceptualizing technological innovation as a process of network building.

Thus, instead of considering technology as an exogenous factor to the system of commodity production, the network approach does not differentiate between human and nonhuman elements. In such analyses, society does not transform technology, nor does technology cause the transformation of society. Rather, "the very actor-networks ... simultaneously give rise to society and to technology."

Moreover, Arunas Juska and Lawrence Busch find ANT to be particularly helpful in uncovering points both of analytical entry and of political action in cases where, as with ABTs, processes of network formation, extension, and reconfiguration are dynamic and in flux. In this perspective, "it becomes evident that the relationships mediated through the network are contingent in nature: they can be disrupted; they can collapse; they can be organized according to different principles; they can be constantly changed and renegotiated".

In a second paper, Busch and Keiko Tanaka deploy the concept of symmetry, which ANT extends to nonhumans, to give "an explicit portrayal of non-human actors over time and space" in "the complex networks known ... as commodity chains", such as rapeseed (canola). The product grades and standards constitutive of commodities represent "rites of passage" for both nature and people.

"Thus, by transforming nonhumans and subjecting them to multiple rites of passage, we co-produce nature, society,

[and] the capitalist market". The symmetry of these qualification processes also extends to their ethical dimensions, which infuse these modes or "rites" of mutual socialization of humans and nonhumans.

In a later analysis of the globalization of rapeseed networks, Busch and Juska contend that the political economy of agriculture perspective and its "categorical apparatus" are "inadequate" to the task at hand. The political economy approach is underspecified because "in true Baconian style, nature is recast as resources to be transformed", as passive. In contrast, the political economy of actor networks reveals how the strategic positions of human and nonhuman actors and the geometry of power in networks are negotiated, modified, and transformed temporally and spatially.

A more recent paper investigates these questions in finer historical detail by examining the reciprocal and contingent relationships between agricultural research and agricultural production, and the changing strategic positions of nation-states as the rapeseed sector has become more globalized since the mid-1970s.

This brief review of rapeseed technoscience provides some indication of the purview and analytical purchase of ANT. With its symmetrical ontology and method, ANT offers a penetrating conceptual repertoire for the analysis of ABTs. Lively eco-social coproductions and category fusions of the natural and artifactual are particularly in evidence in the everyday practices of agro-food networks in an era of genetically engineered organisms.

Thus Bruno Latour extends an invitation to reject the purified categories of "society" and "nature" and focus instead on "the blind spot where society and matter exchange properties". These transactions are brought centre stage in order to expose the inescapable sociomateriality of the entities mobilized into the heterogeneous associations that hold society together.

The constant interchange of human and nonhuman properties in network formation has created "mixtures between two entirely new types of being, hybrids of nature and culture". These processes of mediation, in turn, constitute the foundation of modern technosciences, which "multiply the non-humans enrolled in the manufacturing of collectives and ... make the community that we form with these beings a more intimate one".

Rather than engage in a lengthy exposition of ANT, its merits and limitations, the point to emphasize is that we have a nondualist, relational, and processual framework in which nonhumans are actively present, performative, and consequential. In this significant respect, although otherwise confessedly modest in its claims, ANT challenges the silences and abstractions of production-centered analyses. That is, it furnishes an ontology and conceptual language with which to address the implosions, natureculture hybrids, and newly socialized intermediaries emerging from the heterogeneous engineering practices of agro-food technoscience.

ANT's terms of engagement and insistence on "the permeable boundary running between humans and non-humans" resonate strongly with the ethical and relational concerns of biopolitical activism. An ethical standpoint, a relational moral philosophy, is discernible in ANT's framing of the construction of our world, its insistent reference to crossovers and the exchange of properties between human and nonhuman entities. As Jonathan Murdoch notes, actor-network theorists "force us to look afresh at the categories, divisions and boundaries that frequently divert our attention away from the nonhuman multitudes which make up our world".

This imperative recalls Latour's analysis of Louis Pasteur's research on the anthrax bacillus. That is, the bacilli, once "translated" from the natural competition found in the farmyard to Pasteur's labouratory in Paris, encounter a new environment, an altered state, where they can thrive. Following Latour, Elizabeth Bird observes that "In the labouratory, nature—in the form of Pasteur's "natural-technical object"—becomes an actor negotiating a *new reality*. In the terms of that context, the microbes become actors in shaping a new environment" (258, original emphasis).

Latour continues, "Training microbes and domesticating them is a craft.... Once these skills have accumulated inside labouratories, many cross-overs occur that had no reason to occur anywhere else before". Similarly, when analyzing Pasteur's discovery of the microorganism responsible for lactic fermentation, Latour draws attention to the change in ontological status that this step involved and observes that "in his labouratory in Lille Pasteur is *designing* an *actor*".

The analytical parallels and possibilities of extending this approach to the transgenic "natural-technical objects" of ABTs—such as Roundup Ready soybeans and Bt corn, not to mention the unintended progeny of horizontal gene transfer—are clear. In a brief vignette on ANT, I suggest that agri-biotechnological innovation can be analyzed in terms of the practices deployed to reconfigure or "translate" existing agro-food networks by enrolling genetically engineered organisms (GEOs) and foods.

In effect, corporate networks, such as Aventis, DuPont, and Syngenta, are seeking to displace previous socionatural orderings and to realign agro-food networks in ways that support and "naturalize" the diffusion of GEOs into rural environments, crops, and animals. Thus, "agri-biotechnologies introduce new mediators into the intimate corporeal relations of agro-food networks, promising new corporealities and, quite literally, new bodies".

Biopolitical mobilization, notably in Western Europe, has raised material and ethical concerns against this reordering of agro-food networks in marshaling opposition to the environmental release of GEOs and the incorporation of genetically engineered foods into human bodies. This mobilization directly challenges the industrial, technoscientific problematization of ABTs, variously framed as the answer to world hunger, as an improvement on nature, or in terms of the inevitability of technical change.

In this modernist instrumental perspective, nature is objectified as a field of resources awaiting domination and exploitation by the relentless advance of technoscience. Against this dualist rationality, green biopolitical activism is informed by a relational ethics and by precepts of shared community. These biopolitics of ABTs reveal the "clash of divergent ontologies" provoked by struggles to realign agro-food networks into new socionatural orderings.

By undermining the modernist ideology of nature as externalized and objectified, ANT provides theoretical resources to address nature and its lively materiality directly. This attention to how "socionatures" are constructed broadens critical engagement with capitalist technoscience and political economy and informs our understanding of the heterogeneous associations fostered by this ordering of the socioecological. The dimensions of this political space are amplified if we interpret ANT as an ethical discourse of how to live in the world. This perception of ethical, as well as material, embeddedness speaks directly to the problematics of biopolitical activism, as noted above.

Even a cursory understanding of ABTs brings recognition of the new socionatural assemblages emerging, under the aegis of capitalist technoscience, to construct new worlds. In turn, this recognition suggests the need for conceptual frameworks that explicitly bring normative judgment and political critique to bear on these new collectives and respond to Donna Haraway's interrogation of technoscience: how, for whom, and at what cost?

Acknowledgment of our partnership in these human-nonhuman assemblages would be an initial step toward the development of forms of social organization that encourage democratic choice between alternative orderings and the worlds they bring forth. For these purposes, we need ontologies that reveal, not abstract, our interactive, relational production of worlds we inhabit with others. The novel socionatural assemblages of capitalist agro-food technoscience, which herald new actors and new realities, underline the significance of theoretical and political choice.

In this re-encounter with agro-food studies research on ABTs, the theoretical and thematic continuities of this literature emerge insistently. Indeed, in theoretical terms, there is a powerful sense of deja vu, of involution even, and alternative perspectives remain firmly on the margins. In the preceding sections of this chapter, I emphasized both the primacy of the labour process commodification approach and its ontological limitations for comprehending the new socionatures constituted by ABTs.

This framework constrains efforts to find common ground with the relational ontology and moral economy that inform the biopolitics of environmental movements and Green activists. Such serious limitations, in short, lend urgency to theoretical renewal in agro-food studies and, more generally, to the "greening" projects and explorations of Red-Green rapprochement now under way in various fields of critical social theory.

These projects unmask the political agendas and instrumentalist ethics imbricated in modernist ontological antinomies and their reification. At the risk of repetition, the agro-food studies literature on ABTs has as its analytical focus the ensemble of institutional processes and social relations that have led to the commodification of nature or, as Yoxen has it, to "capitalizing life".

This labour process conceptualization, with its emphasis on the subsumption of nature and its manipulation as a productive force, sees this transition exclusively from the standpoint of the social. Nature is subsumed by purposive social agency, whose dynamic is to be found in the laws of accumulation and social relations of the capitalist mode of production. There is no place here for the relational materiality of nature, its liveliness, or its "boomerang" qualities. This framework does not entertain either notions of natural-social co-productions or the consequences of these assemblages for entities with whom we share this world.

The labour process perspective, in short, does not lend itself to an assured engagement with the new constituencies of agro-food biopolitics. However, "this is not to dismiss the strengths of this perspective. Rather, it is ... to observe that this theoretical lens or 'framing' device does not focus directly, for example, on the new socionatural relations, interspecies metabolisms and exotic corporealities unleashed by agricultural biotechnologies.... these new constellations or assemblages of nature-society relations are key catalytic elements of bio-political activism in agro-food networks".

This review also has drawn attention to thematic continuities, although clearly the scale and intensity of social mobilization have grown sharply with the accelerating deployment of ABTs. These thematic links are amply reflected in this volume. Issues of governance, although not known as such in the 1980s, form a recurrent body of concern. Two themes, already identified previously, have retained particular salience.

The first theme concerns the ways in which ABTs heighten the concentration of industrial control in agro-food systems—"a global oligopoly" in William Boyd's estimation. Exploring the "deep structures of monopoly" within a commodification of nature framework, Boyd extends the analysis of ABTs into the era of life sciences multinationals, genomics, and the competitive imperative to capture value by integrating vertically from proprietary intellectual-property platforms to seed marketing and contract farming.

A social constructivist-commodification perspective also frames the contributions of Scott Prudham and Dennis Kelso to this collection. Following Richard White, Prudham adopts the metaphor of nature as an organic machine to describe the transformation of living organisms into technologies and commodities. Within this framework, he is particularly concerned to track the harnessing of public science to private innovation in the development of forestry biotechnology, the concomitant restructuring of university-industry-state relations, and how this trajectory is shaped by the specificity of recalcitrant nature.

In warning against technological fetishism, Prudham insists that biotechnologies be seen not as things but as bundles of social relations with historical lineages in order to emphasize that technological change is socially produced. Although Kelso's analysis of the deployment of biotechnology in commercial salmon aquaculture broadly fits with the labour process-agrarian transition problematic of agro-food studies, his main concern is to reject its technological determinist inflections.

This endeavor brings Kelso into closer engagement with issues that have been strangely muted in this literature until recently and especially with the sources, forms, and resources of social resistance to biotechnological innovation. Kelso's account of transgenie salmon embraces not only mobilization around questions of food safety and environmental risk but also the defensive stance of salmon farmers against the perceived technological threat to the aquaculture industry as presently constituted.

In elabourating these questions, Kelso draws attention to the politics of scientific uncertainty and state regulation, highlighting the strategic importance of discursive struggles to form public perceptions of nature and the natural. This discussion, together with the chapters by Julie Guthman, Frederick Buttel, and Rachel Schurman and William Munro, begins to address important lacunae in the agro-food studies literature.

The second thematic grouping is around governance issues and focuses on continued First World-Third World tensions over access to genetic resources and the perceived asymmetries of power articulated by intellectual property rights regimes and, notably, the 1994 WTO TRIPS accord. These ongoing tensions find expression in the contested politics, shifting alliances, and arcane regulatory processes of new multilateral institutions of global governance—the WTO and CBD—and their disputed mandates in the conjoined policy arenas of international economics, trade, and environment.

Environmental governance emerges in another contemporary guise in the chapter by Astrid Scholz. Her analysis traces the vagaries of the utilitarian rationale for biodiversity conservation as the importance of natural product screening in the R&D strategies of transnational pharmaceutical corporation's waxes and wanes, exacerbating the inequities of power embedded in private bioprospecting agreements.

The contribution by Frederick Buttel similarly can be situated within the global governance thematic; however, like Kelso's chapter, it also provides a bridge to some neglected issues. Buttel's wide-ranging account encompasses the institutional architecture of the globalization regime, the power brokering of the GATT-WTO transition and the "spoiling" role of agro-food issues—festering EU-U. S. "food wars" (bananas, beef, GEOs), unilateral trade sanctions, the "environmentalization" of ABTs, the global farm crisis—in revealing potential fault lines and pressure points in this accumulation regime.

Taken in the aggregate, these individually contentious issues have cumulatively resonated with growing force in political, cultural, and institutional domains. Whether or not GEOs prove to be the Achilles' heel of the globalization regime, Buttel is surely right to stress that EU-U. S. disputes, and their potential to galvanize social protest, are less about trade liberalization than about the perceived threat of institutional convergence, and especially the forced harmonization of national regulatory structures and the further erosion of cultural identity.

The bridge in Buttel's account rests on recognition of the crucial nexus formed by the politics of scientific research and policy, environmental risk, food safety, regulation, and consumption. These politics are complexly intertwined and varied in their manifestation, ranging from agro-food movements, environmental coalitions, and many forms of direct action, including street theater and green sabotage, to NGO involvement in regulatory processes and consumer pressure on food retailers and manufacturers.

This terrain remains largely unexplored in agro-food studies, apart from limited incursions into one or other of its arenas, such as the politics of rBGH consumption or the role of ABTs in facilitating transition to demand-driven food systems. This neglect is surprising in view of the considerable prominence of these research themes in other fields of critical social theory. In this context, work on the institutional matrix of science, science policy, and regulation, as undertaken more generally by Yoxen, Kay, Bud, and Wright, could fruitfully be extended to developments in agriculture and food since the 1980s.

In the present volume, the chapters by Guthman and by Schurman and Munro set some markers to follow in addressing the research nexus identified above. Guthman examines the contradictions that GEO labeling and right-to-know legislation in the United States present for movements seeking to build an effective politics of consumption, especially the privatization of risk management for genetically engineered foods and the political disarticulation that labeling implies. In calling for research on "an emerging political economy of risk," Guthman follows Haraway in emphasizing the centrality of struggles for "the power to define what counts as technical or political".

In this respect, it would be interesting to explore how European consumer groups and environmental organizations have avoided the pitfalls of labeling and convinced national governments and the EU to redraw the boundary between the technical and the political by agreeing to reopen previous regulatory decisions to public scrutiny. Demarcation struggles and boundary changes are the central theme of Schurman and Munro's chapter. The chronology of events in the rising social opposition to ABTs is becoming well-known, but there is a dearth of careful analyses of antibiotech activism, its strategies and modus operandi. Schurman and Munro begin to address this hiatus by examining antibiotech movements in the North, primarily in the United States.

In their view, the marked shift in public sentiment against genetically engineered foods is not attributable to inchoate consumer resistance or to spontaneous protest. Rather, this dramatic change, and its economic and regulatory repercussions, have been orchestrated by the social-movement organization and mobilization of established civil-society actors.

In pressing this argument, the authors provide valuable insights into the wide array of social organizations articulated by this movement, its tactical sophistication, and its networking skills, which facilitate operation across a variety of institutional and spatial scales and in different regulatory spheres and discursive arenas. Schurman and Munro leave unanswered the thorny questions of representation and the transformative potential of the antibiotechnology movement. However, it is to be hoped that this initial exploration will encourage others to join them in investigating this serious gap in the politics of agricultural technoscience.

Fig. Arabidopsis thaliana

In December 2000, completion of the plant genome sequence of thale cress (*Arabidopsis thaliana*) was announced, the first of some 250,000 plant species. Agro-food technoscience, here represented by the public-private consortium, including Monsanto, which comprises the Multinational Coordinated *Arabidopsis thaliana* Genome Research Project, goes marching on. Now, as in the 1980s, it is vital that critical social scientists join with the antibiotech movement and other progressive forces in the struggle to submit this enterprise and the worlds it would create to democratic debate and public assent.

4

Evolving Policies of Biotechnology

LABORATORY TO FIELD

The timely development and rational introduction of R-DNA modified organisms into the environment depend on the formulation of sound regulatory policy that stimulates innovation without compromising good environmental management. Ecologists are unable to predict which introduced species will become established and which will not, and it is often not possible to explain successes or failure after the fact.

The weighty publicity over biotech products and research directions helped bring about a new configuration of public advocacy. Transgenic animals were an issue that linked animal rights organizations, environmentalists, and alternative agriculture groups. The development of more refined genetic screening techniques brought warnings from civil liberties and disability rights advocates. Disclosures of rising expenditures in the Department of Defense's biological defence programme stimulated interest from activists in the disarmament community.

Feminist-health advocacy groups began exploring the impacts of genetic techniques on reproductive technologies. Bioethicists, clinicians, and religious leaders began tackling the thorny problems of human gene therapy. The prospect of major pharmaceutical advances through rDNA research provided the grist for debates in the international health community on the priorities for developing vaccines. Food and agricultural organizations questioned the impact of biotechnology and the new patent provisions on control over plant genetic resources.

It appeared that every major industrial innovation in applied genetics tapped a wellspring of new issues that were brought to the social agenda. Of these, the issue that ignited the strongest public reaction during the early stages of the biotechnology revolution was the introduction of genetically engineered organisms (GEOs) into the environment. This chapter examines the origins of regulatory oversight over deliberate releases and the federal efforts at creating an orderly transition from small-scale labouratory applications of gene-splicing to large-scale releases of GEOs into the environment.

SOCIETAL CONCERN OVER GEOS

Why has there been so much concern over the risks of releasing plants and microorganisms that have been modified by genetic engineering techniques? Shouldn't the emphasis be on the product and not the particular way the product is created? There are three plausible explanations for the cultural selection of genetically engineered products as a special area of concern. First, the perception of risk associated with deliberate release has largely been formed from prior concerns about recombinant DNA research.

In other words, the environmental problems of genetically modified organisms were inherited from earlier stages of the genetics debates. Recombinant DNA-produced organisms are what R. E. Kasperson calls a "social amplifier" in the public's perception of risk. A second explanation is based upon the notion that genetic engineering provides far greater specificity and control over the product than one could achieve by plant hybridization or breeding of animals. As a result of the specificity of rDNA techniques and their capability of joining quite distant forms of life, the novel life-forms span wider species boundaries and are subject to fewer natural constraints.

By inserting a single foreign gene, a phenotypic property of a bacterium may be radically altered. Resistance to antibiotics is such a property. The ability to change phenotypes in the labouratory with such ease has heightened concerns about deliberate release. Will nature have an opportunity to accommodate to these sudden changes?

To the contrary, some scientists have argued that the specificity of genetic modification makes modern gene-splicing safer than conventional genetics, where genes get mixed randomly and in large clusters. Following this line of reasoning, the precision of gene-splicing means that the resulting properties of the organism will be easier to predict. According to plant geneticist Winston Brill writing in *Science,* in conventional breeding it is impossible to predict the properties of the progeny from most of the crosses.

Genetically engineered plants have greater specificity: "If we compare plants derived from breeding programmes with those derived through genetic engineering, it is clear that, in the latter case, the addition of a few characterized genes to the plant results in properties that are relatively easy to predict."

From another perspective, despite the specificity of rDNA techniques, with them one might be capable of producing more substantial changes in organisms with fewer genetic alterations than with classical genetic techniques. The issue of whether modern genetic engineering techniques are capable of producing varieties of plants, microbes, and animals that could never have arisen from the natural rearrangement of genes remains unresolved. It is widely acknowledged that we humans can at least accelerate or redirect the evolutionary process even if we cannot create qualitatively new life forms.

A third reason why genetically engineered plants and organisms designed for environmental release have attracted more concern than the release of similar products prepared through conventional genetics is related to the reputed power of the new technology. Gene-splicing has been the raison d'être of a technological revolution. This is not simply another discovery in the slow, incremental growth of science. This discovery has given birth to a new industrial process for radically reconfiguring biological matter. The disclosure that there is a new power to transform nature is one of the sources of public and scientific anxiety.

It might be argued that if rDNA technology embodies a power to stimulate the growth of a multibillion dollar industry, why should its risks be considered comparable to those of conventional genetics? What is the likelihood that the industrial potential of gene-splicing (gentechnics), which is, let us say, a thousand times greater than that of conventional biotechnology, will be unleashed without any increases in environmental risks? It is certainly a question worth considering.

Setting aside for the moment the body of scientific argument about the potential risks of new biotechnologies, there is an undisputed equation between technological power and risk anxiety that must be considered in fully understanding the public reaction to biotechnology. The simultaneous pronouncements about power and safety seem incongruous to a popular culture that has been sensitized to technological failure.

In trying to comprehend the risks of releasing genetically altered species into the environment, inevitably we are drawn to comparisons. Two technologies of commensurate transforming quality to gene-splicing are synthetic organic chemistry and nuclear physics. Both of these technologies are capable of creating new arrangements of matter in a fashion analogous to the creation of novel species through biogenetic engineering.

Risk assessment for synthetic chemicals has been in progress for several decades. There have been some important breakthroughs as well as notable impediments. The identification of a chemical substance is a well-defined process.

It is, therefore, possible to construct a precise inventory of chemical compounds. The same is not true for biological agents, at least in some practical sense. Microorganisms and plants are classified by phenotype, and therefore the addition or deletion of a few genes will not necessarily warrant a change in the classification.

It is estimated that 60 to 80 thousand distinct chemicals are used in industry out of a pool of several million that have been synthesized. If a genetic identification system was used for biological organisms, the number of extant chemicals would pale against the number of distinct life-forms since, for the latter, a single nucleotide change would be a differentiating factor. Keeping track of novel organisms and establishing an identification system is a problem of enormous complexity, and probably unrealistic since genetic mutation is a constant occurrence. Yet any serious regulatory effort in biotechnology must address the identity question.

An obvious difference between inert chemicals and life-forms is that the latter are self-reproducing. Throughout the history of the chemical industry there have been countless cases where toxic carcinogenic chemicals were disposed of in lagoons and landfills. These chemicals saturate the soil and eventually migrate to subsurface water supplies where they contaminate drinking water. Once embedded in the earth, many industrial chemicals are difficult to remove.

Entire neighborhoods in areas such as Times Beach in Missouri and Love Canal in New York have been evacuated because of toxic waste. In other, more manageable situations, contaminated soil is removed or filtration methods are applied to poisoned wells.

The mistaken release of a nuisance biological agent cannot be handled by techniques developed for chemical contamination. At the worst, the released organisms are beyond recall and will grow to population orders of a magnitude beyond the density of the inoculation. Moreover, if a novel organism were introduced and subsequently found to be dangerous, geographical isolation and community evacuation would simply not work.

Considerable progress has been made in standardizing toxicological testing for chemicals. The use of accepted methodologies, standardized target species such as germ-free mice or rats, and microbial assays such as the Ames Test have contributed to uniform standards of risk analysis. Notwithstanding the progress, there are still many areas of uncertainty and scientific debate.

Among them are questions about dosage and extrapolation from animal to human effects. Also, human epidemiological studies are frequently too insensitive to pick up small increases in cancer incidence over a lifetime exposure. While there are many effects of chemical exposure that are not well understood, there is at least a basic methodology for gathering the data.

There is no commensurate methodology for assessing the risks of released organisms. Moreover, the risks associated with certain chemical releases are real. Their effects on humans and the biotic world have been observed. In contrast, the potential risks of genetically altered life-forms are currently speculative. As a result, the social demand for evaluating the risks of bioengineered products designed for environmental use may not evoke the same urgency as if the hazards were confirmed.

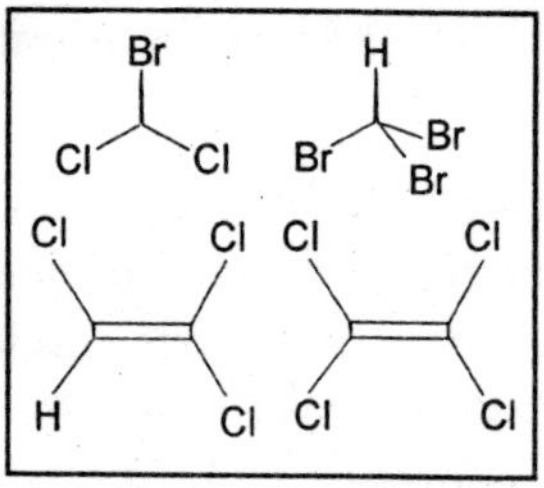

Fig. Trihalomethanes

When chemicals enter the environment, it is not always obvious what effects the breakdown products (metabolites) will have on the ecology. Chlorine has been used extensively to purify drinking water. Its use has been associated with the appearance of chloroform, a potent carcinogen, and other troublesome compounds called trihalomethanes. While there may be risks in the continued use of chlorine, the alternatives are not good. No safer method for purifying water is available. We can never be sure how released chemicals will reconfigure themselves in ecosystems, but compared to the possibilities of biological entities, the range of unexpected outcomes for inert chemicals is probably much narrower since the biological entities mutate.

A subclass of all possible mutations affects the phenotype of the entity. Whether it is more complicated to predict the mutational possibilities of a novel organism than to predict the synthetic pathways and metabolites of a new chemical is the subject of debate. Organisms are certainly more complex than inert chemical compounds. That does not mean that they are inherently riskier, but it does portend a high level of complexity in analyzing environmental risks of genetically novel entities in contrast to newly synthesized chemicals.

There are also some comparisons to be made between biotechnology and nuclear technology. The number of radio nuclides is relatively small, surely in comparison to industrial chemicals. They are generally used in well-defined and controlled settings. There are laws regulating the release of radioactive materials. Also, the health effect of radionuclides in high and moderate doses has been studied and is reasonably well understood. This is not the case, however, for low-dose, long-term exposure.

Radioactive materials are detectible in minute quantities with a sensitive monitoring device. To improve the confidence of residents living in the vicinity of a nuclear power plant, some communities have been provided with radiation detection counters.

There is nothing analogous in biotechnology. In theory, one can identify and track bacteria that have been released into the environment. The organisms would have to be tagged in a special way. Even then the identification can be a difficult task depending upon the behaviour of the microbes and the conditions of the environment. Each case is unique. At this time, there are no standard methods of detection and no cannonical procedures for distinguishing safe from deleterious organisms. In comparison to bacteria, it is much simpler to detect the spread of an unwanted plant (a weed). But once released, plants, like bacteria, may be impossible to recall.

We have seen how, within a period of a decade, a single critical discovery was the progenitor of an industrial revolution. The investment, scientific, and corporate communities moved expeditiously to capitalize on the commercial opportunities of the new genetics. To whom was this new industry accountable? How were the public policy issues handled? What social guidance was imposed upon the new technological direction? The next section discusses the early regulatory response to environmental applications of biotechnology.

NIH'S EARLY ROLE

Beginning in the 1980s, industry and university proposals for field-testing genetically modified plants and organisms triggered a major science policy debate in the United States that spilled over to the European community. Those who have followed the recombinant DNA controversy from its inception will recognize that the current configuration of policy alternatives is the result of a ten year historical process. Initially, molecular geneticists cast the problem of genetic engineering in technological terms. Gradually, public perception of the problems associated with gene-splicing focused attention on the ethical and ecological issues.

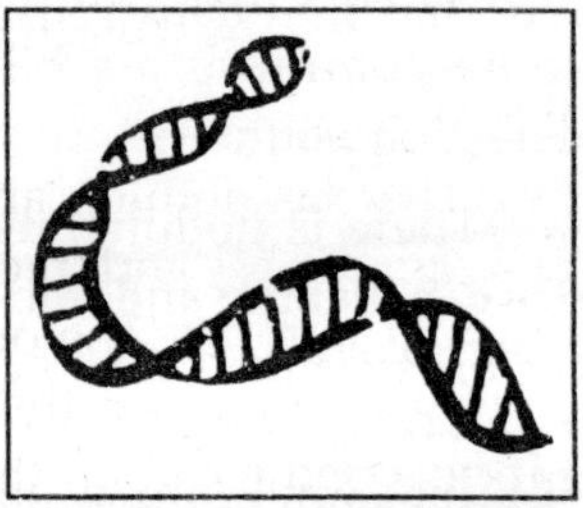

Fig. rDNA

The emergence of a second generation of genetics policy debates brought participation from new disciplines, new communities, new public interest groups, and new federal agencies. Public concerns slowly shifted from the singular issue of labouratory safety to a much broader range of problems. And while these changes were taking place, regulatory oversight of biotechnology also shifted from the National Institutes of Health to other governmental bodies.

For nearly a decade, the agency that assumed primary responsibility for the safe uses of genetic engineering was the National Institutes of Health (NIH). Essentially a science-funding agency under the Department of Health and Human Services, the NIH established the Recombinant DNA Advisory Committee (RAC) in 1974 at the recommendation of Paul Berg, the Stanford University biologist who provided leadership in the early efforts to assess the risks of rDNA research.

According to Berg's plan, an international meeting of biologists would result in a set of principles for safe handling of rDNA molecules. Those principles were then to be used by the NIH's newly formed scientific advisory committee to establish guidelines for all genetic experiments involving the cutting and splicing of foreign genetic material. Berg, along with other scientists who organized the Asilomar meeting, surmised that if the NIH did not act with dispatch in responding to the potential risks of gene-splicing, Congress might pass restrictive legislation. They viewed the passage of such legislation as detrimental to the interests of biology; particularly at stake was the legitimacy and progress of molecular genetics.

First, biologists could lose influence over the risk assessment and risk management process. Second, the field of molecular genetics might become stigmatized as the only discipline whose principal research method was regulated. Third, there was considerable concern that rDNA legislation would be inflexible and difficult to amend. Asilomar organizers feared that biology would be saddled with irrational requirements.

Written by scientists for the general use of scientists, the NIH guidelines made their debut in June 1976. No explicit references were made to industrial processes or non-NIH supported uses of rDNA. The guiding principle behind the development of the guidelines was containment. Since one could not predict at the outset with any reasonable degree of certitude that a foreign gene introduced into an organism could not inadvertently transform it into an epidemic pathogen, the consensus at Asilomar was to construct a set of containment provisions consisting of physical barriers, safety operations, and carefully selected host organisms chosen because they do not survive well outside of the labouratory setting.

Each containment level was matched with a class of experiments that was permissible under the stipulated conditions. As scientists on the RAC and elsewhere became more confident in the safety of rDNA techniques, labouratory containment requirements were substantially relaxed.

The early NIH guidelines contained a provision that restricted industrial applications of rDNA technology. Cultures of rDNA organisms produced or handled in volumes greater than ten liters (classified as large-scale) were prohibited for NIH grant recipients. The large-scale prohibition was not based upon a scientific assessment of risk. It was a convenient threshold introduced by academic scientists to protect the use of standard labouratory beakers in basic research.

A second provision of the NIH guidelines restricted industrial activity by explicitly proscribing the intentional release of an rDNA organism into the environment. Since absolute containment was nothing more than an idealization, unintentional releases were considered unavoidable. However, by limiting the volume of rDNA culture, the probability of escape could be minimized. Also, since the volume of agent released is correlated to survival and propagation, the large-scale prohibition also supported the general containment strategy.

Commercial interest in rDNA techniques grew rapidly in the mid- 1970s. Estimates of industry growth vary. My own count indicates that a minimum of 14 new biotechnology enterprises (NBEs) were formed in 1976, the year the NIH guidelines were introduced. During 1979, NBEs grew by at least 26; and in 1981, at least 66 NBEs entered the biotechnology industry.

The NIH had no legal jurisdiction over the research in private companies. In practice, however, firms were not willing to risk the negative publicity that might arise if they violated the NIH guidelines. The nascent biotechnology industry was comprised predominantly of small firms started by relatively young scientists, many of whom retained their university affiliation. The close link between academe and industry may help explain the high degree of compliance among new firms with the standards adopted by the NIH.

Since these scientists were groomed on the NIH guidelines, the complications of compliance that might have beset a new industry were minimized. Despite a watchful media, there is no evidence that the NIH guidelines were flaunted by the biotechnology industry. Ironically, the few cases where violations of the guidelines were reported took place at universities.

Geoffrey Karny cited two factors responsible for industry's compliance with the voluntary guidelines: "First, the possibility of tort lawsuits provides monetary inducement to comply with the *Guidelines,* which would probably be accepted as the standard of care against which alleged negligence would be evaluated. Second, the threat of statutory regulations, which the companies have sought to avoid, always exists."

Between 1976 and 1979 the NIH process for overseeing rDNA research was put to its severest test. First, the city of Cambridge, Massachusetts, issued a moratorium on rDNA experiments requiring moderately high physical and biological containment. After a widely publicized citizen review process, the city passed the country's first rDNA law in 1977. The law departed from the NIH guidelines in a few minor respects. More importantly, it symbolized the right of local government to exercise control over where the research gets sited and the safety conditions of its performance. Moreover, it established uniform and legislatively mandated guidelines for both publicly and privately funded research.

After the Cambridge rDNA law was passed, nearly two dozen states and local communities debated the issues. Legislation was enacted in about half the jurisdictions. In response to local events and a national mood of concern toward gene-splicing, fifteen distinct bills were filed in Congress between 1977 and 1978 to regulate rDNA research. Some of these bills would have shifted the regulatory authority from NIH to a national commission.

These bills also varied in the degree to which local laws were subject to federal preemption. Congress spent two years debating the issue of an rDNA law. A compromise bill was finally voted out of committee early in 1978, but for lack of strong congressional leadership and interest it failed to reach the House or Senate floor for a vote.

Congressional failure to enact legislation strengthened the NIH's position as the sole agency overseeing rDNA activities. Responding to continuing public concern over the research, the Department of Health, Education, and Welfare (HEW, currently the Department of Health and Human Services) rewrote the RAC's charter and increased the size of the committee from sixteen to twentyfive. The new charter, issued in 1978, stipulated that one third of the committee was to consist of individuals with expertise and interest in public health and the environment.

The change in the composition of the RAC drew sharp criticism from prominent scientists who argued that rationality was being compromised by including nonscientists in what was essentially a technical process. When the expanded RAC met in early 1979, its agenda was filled with petitions for relaxing containment requirements and approving additional host-vector systems.

Over the next few years several important changes were made in the rDNA guidelines that established a role for the NIH in the review of field tests. A voluntary compliance programme was established that gave the private sector access to the NIH review process, while prohibitions against large-scale rDNA activities and the intentional release of genetically altered strains into the environment were removed.

After the decision by Congress not to enact rDNA legislation, there was a strong residue of criticism that the NIH guidelines could not protect society from the potential adverse consequences of commercial gene-splicing. In response, NIH developed a voluntary compliance programme for institutions that did not fall under the agency's purview. This initiative gave the fledgling biotechnology industry the opportunity to gain the imprimatur from the NIH for both labouratory and commercial-scale rDNA work.

A firm wishing to participate in the programme first submitted the composition of its institutional biosafety committee (IBC) to the NIH's Office of Recombinant DNA Activities for approval. Once IBC approval came, a firm could file requests with the RAC following procedures similar to those of university petitioners. One difference between the NIH's handling of academic and industry proposals is that, on the occasion of the latter, the RAC went into closed session.

Members were required to sign confidentiality pledges for the protection of information deemed proprietary by the firm. Some RAC members were opposed to having the committee review proposals in closed session. One public interest member refused to participate in the review of industry proposals. He argued that oversight of the private sector was the responsibility of those agencies of government with statutory authority to protect workers, public health, and the environment.

Since the NIH lacked authority to carry out these functions, he felt his participation would give legitimacy to this extension of the NIH's role. Another RAC member expressed the following sentiment: "Voluntary compliance is the worst of all possible worlds. You achieve none of the objectives of regulation and none of the benefits of being unregulated. All you're saying is 'I give a stamp of approval to what I see before me without any authority to do anything.' "

The programme was also the target of mainstream critiques. As an example, in its 1981 report on biotechnology, the Office of Technology Assessment wrote: "The most significant limitation in the scope of the Guidelines is their nonapplicability to industrial research or production on other than a voluntary basis. This lack of legal authority raises concerns not only about compliance but also about NIH's ability to implement a voluntary programme effectively."

In May 1979, the NIH's advisory committee went on record opposing voluntary compliance by a vote of nine to six, with six abstentions. The RAC voted that non-NIH funded institutions should be required to comply with the guidelines. This recommendation notwithstanding, the voluntary compliance programme became a permanent part of the guidelines in January 1980.

Many biotechnology companies considered it in their long-term interest to secure RAC approval for their rDNA activities. Regardless of how the firms' management felt about the potential risks of gene-splicing, submitting experiments for NIH approval was excellent public relations. The voluntary compliance programme helped the biotechnology industry respond to the criticism that the private sector was operating without regulations. When pressed by local communities to demonstrate the safety of genetic experiments, a company's most compelling response was that it complied with the NIH guidelines.

In the wake of sporadic episodes of local opposition to genetic engineering research, the biotechnology industry sought a predictable and stable regulatory climate, but one that would not impede research and development. The NIH contributed to this goal, but with limited success. The voluntary compliance programme proved functional to the industry during its early years of development when there was intense competition, investment instability, and the uncertainties of local regulation. As the commercial activities progressed from labouratory research to product development, the NIH policies accommodated to the new stage of industrial activity, particularly in their response to large-scale work and the release of genetically modified plants and organisms into the environment.

LARGE-SCALE PROHIBITIONS

Provisions were built into the early NIH guidelines for waiving the large-scale prohibition in cases where minimal risks were balanced against important societal benefits. The wording of the prohibition was clarified in the 1978 version of the NIH guidelines. No exceptions to the prohibition against the production of large scale cultures were permitted "unless the recombinant DNAs are rigorously characterized and the absence of harmful sequences established." Under NIH's leadership the risk assessment paradigm was in large measure still under the primary influence of geneticists.

For a short period of time, the NIH gave serious attention to all phases of large-scale work with rDNA molecules. Proposals submitted to the RAC were required to include a description of labouratory practices, specifications on physical and biological containment, risk data, characterization of genes, and the design of the fermentation equipment in conjunction with the physical description of the facility. In 1980, the RAC published a standard that it planned to use for reviewing large-scale proposals.

The committee's role in evaluating plant design and operations for large-scale fermentation drew criticism from some members. They argued that the RAC should confine itself to advising the NIH on the nature of biological procedures and not plant operations. The committee, after all, had no special expertise in bioprocess safety engineering. It depended on outside consultants for guidance in this area. Also, it was a matter of some significance that molecular biologists on the RAC found the review of plant design boring and uninformative.

Other RAC members contended that if the voluntary compliance programme was to mean anything, a comprehensive review of large-scale operations was essential. Since no other federal agency was engaging in this review, they believed that it was the NIH's responsibility to fill the regulatory void.

The internal debates over this issue were a poignant reminder of the NIH's ambivalence in serving as overseer of commercial genetic engineering. These debates also cast doubt on the logic of having a biomedical funding agency guide a nascent industry exclusively through a system of voluntary measures. The peculiar nature of this regulatory programme began to reveal its contradictions. For example, within the RAC opposing factions interchanged positions.

Initially, the group most skeptical about the safety of genetic engineering expressed opposition to the RAC's review of industrial proposals, particularly large-scale ones. They reasoned that the NIH was acting as de facto regulator without enforcement powers or congressionally derived authority. If the RAC refused to serve this function, members of this faction believed that agencies like the Occupational Safety and Health Administration (OSHA) and the Environmental Protection Agency (EPA) would enter the field.

Concurrently, there were other members of the RAC who opposed broader federal involvement in biotechnology, particularly legislative or rule-making actions. They viewed NIH's continued role in overseeing commercial rDNA work as essential to preempting such initiatives.

The unexpected reversal took place among committee members around 1980. The pro-regulatory group grew less sanguine about the involvement of other agencies in regulating biotechnology. As a consequence, this faction began supporting a stronger role for the RAC. Meanwhile, the committee's regulatory minimalists, also confident that broader agency involvement was unlikely, backed an NIH role over commercial rDNA activities that were limited to issues of pure genetics, namely, sequence characterization and approval of host-vector systems.

The regulatory minimalists succeeded in so limiting the RAC's review of the safety of rDNA products. By 1981, firms taking part in the NIH's voluntary compliance programme were no longer obligated to submit data on sterilization of their fermentation system or their procedures for disposal of rDNA cultures.

This policy change was expressed in a proposition adopted at the September 1980 RAC meeting: "The RAC will determine if a given recombinant DNA containing strain is rigorously characterized and the absence of harmful sequences is established. Such a determination shall include specification of a containment level. These determinations should not in any way be construed as RAC certification of safe labouratory procedures for industrial scale-up."

The potential release of rDNA organisms through the effluent of bioreactors did not attract any public attention. While the EPA had jurisdiction over such biological releases under the Clean Water Acts, a constituency for agency action did not develop. The newly formed Committee (later Council) for Responsible Genetics published an article on biogenetic waste in its public interest bulletin *GeneWatch*.

However, the issue of biogenetic waste was not a rallying point among environmentalists or the general public. It seemed to lack some important features that influence risk selection. First, it did not impart a clear and present environmental danger. Second, the waste stream, even if it carried genetically engineered organisms, did not excite the media. Bioeffluent was not intentionally designed to alter nature. Since science writers are captivated by the frontiers of science, there was not much of a story in fermentation sludge.

Quite a different media and public reaction took place when proposals appeared before the RAC requesting approval to field-test genetically modified plants and bacteria. There is an informative distinction to be made in the public reaction between intentional and unintentional releases of genetically modified life-forms. To fully grasp the distinction, we have to look at the kinds of experiments proposed, the types of media coverage they were given, the role of scientists in raising safety concerns, and the development of a public interest constituency.

Deliberate Release of Gems

As the RAC reduced its oversight over industrial bioprocesses, it became more active in reviewing environmental releases of genetically engineered microorganisms (GEMs). Consequently, field-testing was given greater visibility in the media. From 1980 to 1984, when companies and university scientists were preparing to field-test their products, the NIH was the only federal agency with active responsibility over deliberate release experiments involving genetically altered plants or microorganisms.

Progressive relaxation of the rDNA rules finally led to the removal of barriers to field-testing. The first revisions of the guidelines in 1978 still prohibited "deliberate release into the environment of any organism containing recombinant DNA," but individual waivers were permitted after proper public notification, RAC review, and approval by the director of the NIH.

The revised NIH guidelines of 1982 eliminated the entire list of proscribed experiments. By June 1983, the prohibition against intentional release of rDNA organisms was replaced by a multitiered review process. Submissions for deliberate release required approval by the RAC, the institution's biosafety committee (IBC) and the NIH director, in addition to various subcommittees.

Agricultural applications of biotechnology, widely publicized in the media, were being readied for field trials. By December 1983, the RAC had reviewed and approved three proposals for releasing genetically altered life-forms. In each case, the committee concluded that the tests posed no significant risk to health or the environment.

The first three proposals for deliberate release came from university scientists and therefore did not involve proprietary information. In March 1980, a Stanford University scientist requested approval from the RAC to field-test a corn plant into which had been inserted the corn storage protein gene with modified sequences. The genetically altered strain included a new corn protein gene that encoded all the amino acids essential to humans (including an enhancement of lysine and methionine, in which corn is deficient).

By genetically engineering the new corn protein with the full complement of the essential amino acids, its value to the human diet would be improved. The RAC failed to come up with a hazard scenario for the genetically modified corn; nevertheless, the investigators were required to detassel the plant to prevent pollen dispersion during the field trials. Permission for the test was granted in August 1981. There was no significant public reaction to the decision.

A second proposal for field-testing was brought to the RAC by a Cornell University scientist in June 1982. Tomato and tobacco plants were transformed with DNA from yeast and *E. coli* to provide them with antibiotic resistance. In reviewing the experiment, the RAC raised concerns about the possible spread of antibiotic resistance and the effect of an itinerant plant on the ecology. Neither of these concerns delayed the committee's approval, which it gave on October 25, 1982. When the recommendation reached the director of the NIH, he referred the proposal for a reading. After USDA approval was given in February 1983, NIH accepted the field test in April 1983, nearly a year after the proposal was first brought to the RAC.

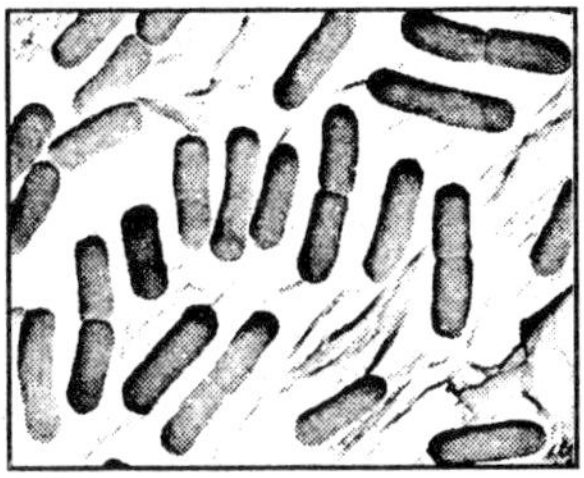

Fig. E.coli

The third of the early field test proposals for genetically modified life-forms proved to be the most controversial. Ironically, it was viewed by some experts as among the safest deliberate release experiments one could perform in the environment. There was one important distinction between the first two proposals and the third: the latter consisted of genetically altering a soil bacterium. The difference between plants and bacteria proved to be a critical factor in the public perception of risk. Also, there was an established tradition of introducing hybridized plants into the environment. However, no analogous tradition existed for microorganisms. The relative novelty of this field test was reflected in the RAC's review.

In September 1982, scientists at the University of California at Berkeley proposed to field-test two soil organisms (*Pseudomonas syringae* and *Erwinia herbicola*) from which about a thousand base pairs of DNA sequences had been deleted. In their natural state these organisms synthesize a protein that provides a nucleation point for ice crystallization, and are known as ice-nucleating agents (INA).

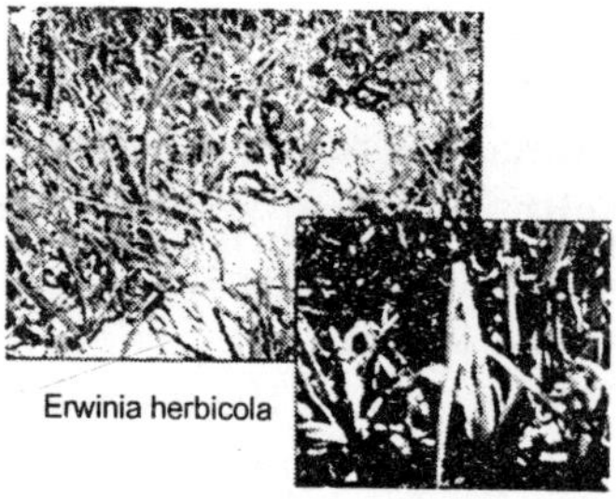

Fig. Erwinia herbicola

By excising the genes responsible for ice nucleation and establishing the genetically modified organism in the environment, scientists believed they could reduce the temperature at which frost begins to form on crops.

Fig. Pseudomonas syringae

When the proposal for field-testing the microbes with the deleted ice-crystallization gene (denoted INA^- or ice minus) was first brought before the RAC, one member expressed concern about the role the organisms might play in the environment. Questions raised about the effect of INA^- on precipitation patterns were based upon some theories that INA^+ (normal ice-nucleating agents) played a role in atmospheric weather. The precipitation issue was not resolved by the committee. Nevertheless, the RAC approved the field test, although by a small plurality (seven yes; five no; two abstentions).

When the recommendation was transmitted to the director of the NIH, more data along with additional safeguards were requested. Several months later on the second round of review, the RAC approved the proposal unanimously. The NIH gave the final go-ahead for the field test in June 1983, some nine months after it had initially received the proposal. Investigators agreed to mark the strains of the ice minus organism with antibiotic resistance, which allowed them to monitor the dispersal of the organism.

They also agreed to limit field-testing to a single location, the University of California agricultural field station at Tulelake in northern California, a site isolated from the major fruit tree and citrus-growing regions of the state. Despite these precautions, the RAC's decision on the field test for ice minus was met with citizen opposition and litigation from the Foundation on Economic Trends, Jeremy Rifkin's organization based in Washington, D.C. Between 1980 and 1985 the RAC reviewed five proposals for field-testing GEMs. Four were approved and one was voluntarily withdrawn.

GENETICS AND THE ENVIRONMENTALISTS

In the mid- 1970s, when rDNA molecule research was still an exotic technique, the public entry into the debate over its safe uses was determined by several factors. First and foremost, a group of distinguished biologists called attention to the new technique by publishing letters in science journals that caught the attention of the popular press. The international meeting at Asilomar was attended by science writers, whose coverage of the events dramatized the disagreements among biologists at the meeting.

In drawing attention to the new discovery, the Asilomar organizers did not intend for the fate of the new technology to be decided by popular acclamation or democratic process. However, some biologists, frustrated by the NIH's role in setting safety standards, brought their concerns to university campuses and local communities. To the public, it appeared that scientists were polarized on the dangers of rDNA research.

Second, the issues were dramatized through the world press by the controversy that erupted in Cambridge, Massachusetts, which pitted community values against the interests of national science. The 1976 Cambridge rDNA controversy represented the birth of public involvement in genetic engineering. Prior to that, the issues were debated exclusively in professional groups and on university campuses, notably the University of Michigan.

Third, national environmental groups such as Friends of the Earth, the Environmental Defence Fund, and the Sierra Club devoted some attention to genetics issues between 1976 and 1979. But the commitment by these groups was limited usually to a single staff person. Their activities were met with skepticism, and in some cases, disapproval by some board members who had ties to the biomedical community. Fourth, but to a lesser extent, litigation had drawn public attention to the issues. A resident of Frederick, petitioned the court to enjoin the NIH from performing a risk assessment experiment until an environmental impact statement was prepared. A second suit filed by Friends of the Earth alleged that HEW failed to comply with the requirements of the National Environmental Policy Act when it established policies for rDNA research.

Fifth, congressional interest in the safety of rDNA research intensified in the mid-to-late 1970s. House and Senate hearings provided a forum for activist scientists and environmental lobbyists. The accumulated effect of these factors resulted in heavy media coverage for both local and national events related to gene-splicing experiments.

The subsequent public debate over the specific issue of the deliberate release of genetically engineered organisms has been fostered by a similar configuration of factors, namely, scientist-critics, litigation, environmental coalitions, local opposition to experiments, and congressional interest. However, the old and new rDNA debates differ in the weighting of these factors.

For example, Jeremy Rifkin has proven particularly effective in drawing public attention to the environmental and ethical issues associated with the deliberate release of GEOs. Operating with the modest resources of his nonprofit organization, Rifkin has utilized a variety of techniques, including confrontational protests, litigation, and short-term coalition building, to dramatize breaking issues in genetics and public policy.

Rifkin's philosophical opposition to genetic engineering has been expressed in his writings and media accounts that feature his work: "Genetic engineering represents the ultimate negation of nature." On genetics policy, he has had more impact on the media that any single group or individual in the United States. His notable success as a publicist and genetics critic can be explained by several factors. Rifkin can act quickly since he is not accountable to a board of overseers or a mass organization. He is able to identify and capitalize on the weakest link of any policy process. With properly timed and targeted litigation, Rifkin has been able to strike at the jugular of the bureaucracy.

He has proven time and again that a well-placed lawsuit is a magnet for media attention. Rifkin made his national debut as a genetics activist in 1977 when he and a group of supporters invaded a meeting at the National Academy of Sciences devoted to rDNA science and policy. His strategy at the NAS is best described by the 1960s term "guerrilla theater," a dramatically staged political confrontation. Fearing major disruption of its conference, academy officials allotted Rifkin time at the start of the programme for a statement. Meanwhile, his associates, donning silk stocking face masks, stood like motionless icons, projecting a grotesque apparition of human mutants. While Rifkin took over the podium of the academy auditorium, his colleagues stretched a banner across the stage with a quote from Adolf Hitler that stated: "We will create the perfect race."

On another occasion, Rifkin organized a signature campaign of scientists and religious leaders who were asked to support a statement against genetic manipulation of the human germline. The list of supporters included individuals of vastly different ideological persuasion. To have such a politically diverse group on the same side of an issue was itself a news event.

In the fall of 1983, when the RAC was reviewing a proposal for field-testing ice minus, Rifkin brought national attention to the problem of deliberate release by filing suit against NIH on the grounds that its approval was given prior to a full study of the environmental consequences. His suit was supported by two organizations, Environmental Action and the Environmental Task Force. To strengthen his case, Rifkin received sworn affidavits from leading ecologists who opposed the release of the ice minus strains, at least until more could be learned about the environmental consequences.

Rifkin's lawsuits have been rallying points for critical debate on genetics issues. Philosophically, he personifies the modern genetics Luddite. But on a practical level, he has made effective use of environmental law to impose a greater burden of responsibility for assessing the risks and eliciting the ethical consequences of new technologies. Ironically, he is secretly applauded by many who, while they might differ with him philosophically, find a certain degree of relief in slowing the pace of regulatory approval.

Environmental groups, skeptical about deliberate release, have been hesitant to act more aggressively because the hazards of genetic engineering are still hypothetical. For Rifkin, it was less important to prove the hazards exist than to rally support against the hubris of science and reexamine its assumptions about progress.

There is a striking comparison between the old and new genetics controversies on the role of scientific participation. In 1976, the molecular geneticists were taken to task for not calling upon infectious disease epidemiologists to evaluate the risks of transforming *E. coli* into an epidemic pathogen. Subsequently, the infectious disease community did participate in the risk evaluation.

Similarly, ecologists played no role in the rDNA debate until the controversy over deliberate release erupted. Their entry into the discussion over deliberate release came when the social and political context was ripe for their participation. Ecologists are viewed as natural allies to environmentalists; when the latter became involved in the debate over field tests, they sought advice from the former.

Also, the Committee on Science and Technology of the House of Representatives made specific recommendations to NIH and USDA that they revise the memberships of their advisory committees to include individuals specifically trained in ecology and environmental sciences.

Finally, the EPA began developing its own policy on deliberate release in the early 1980s. The agency is accustomed to working with ecologists, for example, in evaluating pest control methods. The first genetically engineered organisms planned for environmental release were classified as biological control agents. The EPA began hiring consultants to study the potential environmental problems of releasing such agents.

A terrestrial ecologist from Oak Ridge National Lab oratory, Frances Sharples, completed a study for the EPA in 1982 on the effects of the introduction of organisms with novel genotypes into the environment. Microbial ecologist Martin Alexander of Cornell University chaired an EPA study group on biotechnology. In 1984, ecologists participated in an EPA sponsored workshop at the Banbury Centre at Cold Spring Harbor, New York, on the evolutionary consequences of biotechnology.

Also, in 1985, with support from the EPA, the USDA, national environmental organizations, and corporations, ecologists, geneticists, public interest representatives, and federal regulators met for four days at Cornell's biological field station in Bridgeport, New York, to develop principles of containment for genetically engineered organisms tested under field conditions.

By raising the issues of biotechnology in the national media, Rifkin also helped catalyze the concerns and perspectives of ecologists on the problems of deliberate release, particularly those who had not been brought in as consultants to the major agencies. Eventually, letters from ecologists appeared in *Science,* followed by responses from geneticists. A dialogue was opened between molecular geneticists, microbiologists, and ecologists on their approaches to risk assessment. Soon it became evident that the issues over ice minus unfolded into a larger debate about scientific culture, epistemology, and disciplinary hegemony.

We will see how the ice minus bacterium became an international symbol for resistance against biotechnology by some groups like Earth First and the European Greens. To the biotechnology industry, ice minus had quite a different symbolic value as a product that could raise public confidence in a new technological frontier that promised to advance agricultural efficiency while making peace with nature.

5

The Alternative Agriculture

THE NEGLIGENCE

This chapter makes the claim that, despite the emergence of alternatives, the trend towards genetic uniformity is likely to be perpetuated owing to the emphasis now placed upon new biotechniques. Biotechniques provide more powerful tools to achieve genetically uniform ideal plants, but they could also be deployed as part of strategies aimed at promoting deployment of *in situ* diversity. The argument presented here is that biotechniques are closing the door which alternative movements managed to prise open, if only briefly, in the 1980s. This could be made in various ways, but here, the argument is developed through analysis of the intellectual property rights (IPR) issue, since this has radically affected the issue of genetic resource control in recent years.

Support for biotechniques takes place against the backdrop of a global economy in which concern to maintain 'technological leadership' and 'competitiveness' has prevented serious consideration of important questions concerning technological options, and the need to consider the matching of ends with technological means. The more deep-seated this concern, the more likely it becomes that institutional change will underpin it, further legitimising the same concern. This is the domain of Arthur's self-reinforcing mechanisms. For this reason I suggest that we are becoming increasingly locked in to genetic uniformity only moments after it seemed that the door was possibly opening for (again, possibly) genetically more diverse alternatives.

GENETIC RESOURCE CONTROL

When southern corn leaf blight hit the US maize crop in 1970, the solution was found not through re-introducing genetic diversity, but in genes from Mayorbala maize found in a West African field. West African germplasm was used at virtually no cost to save US farmers from a disease that had cost them $1 billion in 1970. That US farmers should have been assisted so freely owed itself to the system of free exchange of germplasm operating at the time, based on the principle that biodiversity was 'the common heritage' of humankind.

The UN's Food and Agriculture Organisation (FAO) had discussed genetic resources at its founding conference in Quebec in 1946, and with the International Biological Programme (IBP) hosted a conference on plant genetics in 1961. Little was being done to address the issue of genetic erosion. Most collecting of germplasm was done by academics in universities, and virtually all was done at a sub-national level, an exception being that conducted on maize. The issue of crop germplasm conservation was highlighted by the work of Erna Bennett, who in 1967 organised the second IBP/FAO Conference on genetic resources and subsequently set up the Crop Ecology Unit at the FAO in Rome.

Only a year after southern leaf blight destroyed much of the US crop, a cold winter followed by a dry spring drastically affected the wheat crop of the Soviet Union. Much of the land was planted to the Besostaja variety which was neither cold tolerant nor drought resistant. Wheat prices shot up, and when famine hit the Sahel region in Africa, Malthusianism became fashionable once more. Against this backdrop, Sir Otto Frankel, Bennett's co-editor for a path-breaking book on the subject, persuaded the 1972 Stockholm Conference on the Human Environment to adopt a resolution calling for concerted action on genetic resources. The need to conserve genetic resources was at last being taken seriously as the advancing BCM mode hastened their disappearance.

Hambridge and Bressman recognised long ago the tensions between free exchange in germplasm, and unfree exchange of the product derived from it:

From its rivals a nation may get the wheat germplasm that enables it to supply its own needs or overwhelm those rivals in international trade ... Will nations have the wisdom to deal with this situation, or will it lead to more bitter rivalries and more deadly conflicts, as the beneficent science of chemistry has enormously increased the deadliness of war? In his use of modern science, man has proved again and again that he is a bright child playing with fire.

The Wardian case, invented in 1829, facilitated an exodus of germplasm to one of a growing number of botanical gardens. The 'botanical chess game' has played an important role in shaping the international division of labour, a fact recognised by Marx:

You perhaps believe, gentlemen, that the production of coffee and sugar is the natural destiny of the West Indies. Two centuries ago, nature, which does not trouble herself about commerce, had planted neither sugarcane nor coffee trees there.

Metropolitan powers appreciated that control over commodity trade depended on restrictions on the movement of germplasm. History is therefore replete with examples of the heroic efforts of plant explorers in overcoming embargoes on the movement of seeds, the breaking of which was in many cases punishable by death.

Only in the post-colonial world did 'free exchange' reach its truly international apogee. Some interpret this as allowing gene-poor northern countries in the developed world to maintain access to germplasm residing in the gene-rich, financially poor, countries of the developing south. Yet if, as Galeano writes, the international division of labour was organised 'not by the Holy Ghost but by men', so it was with the system of germplasm exchange. As such, it could be changed by men.

The Seed Industry

Unless they were crossed inbreds, farmers could save seed for planting the following year without appreciable yield loss. The first claims for plant patents were made in 1885. In 1922, lawyers met in London to discuss patent protection for plant varieties but no action followed. It was the nursery industry which was primarily responsible for the passage of the Plant Patent Act of 1930 in the United States. For nurserymen, the obstacle to proprietary ownership of varieties lay in competition from other nurserymen, not farmers, since trees could easily be propagated in competing nurseries.

The Plant Patent Act of 1930 made it possible for asexually reproduced plants to be patented, with the exceptions of potatoes and Jerusalem artichokes. The rhetoric used in support of the act, that breeding had made such significant advances over the past decades, was actually completely irrelevant as far as asexually reproduced plants were concerned. Most of the work done by nurserymen lay in multiplying varieties that had been discovered by chance, and that were the product of insect or wind pollination, raising issues as to whether they should have been eligible for patents.

Fowler concludes that: 'The PPA did not recognise the individual inventor or the creative act as much as it recognised and rewarded the system that produced the new variety, whether by luck or by designs'.

In France, ever since the turn of the century, rose breeders had been seeking the same recognition as inventors of machines. Early attempts were rejected by lawyers on grounds that even full disclosure would not make it possible for breeders to reproduce a variety. By 1928, however, there existed in the Ministry of Agriculture *de facto* protection of breeders' rights through an 'identity and purity service'.

The Italian High Court declared plant varieties patentable in 1948, but confusion led to calls for a special plant patent law. By 1957, with the view that plants and animals should not be patented in the ascendancy, the International Association of Plant Breeders for the Protection of Plant

Varieties accepted an invitation to host a conference in Paris on plant breeders' rights (PBR), leading to the establishment of the Union for the Protection of New Varieties of Plants (UPOV) in 1961, whose International Convention was revised in 1972, 1978 and 1991.

For the most part, PBR legislation has been true to the UPOV Convention, requiring that plants pass the DUS test. The 1960s and 1970s saw several countries either joining UPOV or implementing a system of PBR of their own. Key to the passage of these acts was the belief that sexually reproduced plants could breed true, which, for sexually reproducing plants, is only the case for pure-line varieties (those having undergone four to nine generations of selfing). Also critical was the definition (thought by many to be impossible) of the term 'variety'.

The UPOV resorted to 'nothing other than a description of the steps of the method of breeding', or more accurately, pure-line (Mendelian) breeding. Hence, the extension of IPR to plants through PBR was an institutional innovation shaped by, and made possible by, changes in breeding techniques and technology respectively. However, these were not institutional changes waiting to be implemented as soon as these techniques emerged. Just as there were technical options open to breeders, so the institutional changes made represented a choice from myriad possibilities.

PBR facilitated an increasingly international outlook on the part of the seed industry. Modern varieties were spreading across the globe, and notwithstanding some efforts to improve disease resistance in new varieties, new seeds made increased use of other inputs more likely. In the 1980s, policy-related or structural adjustment lending undertaken by the World Bank advised privatisation of input supply industries, and an expanded role for private sector seed research and distribution, especially in the development of hybrids. The emphasis began to shift, as it had done in the developed world, away from the public system in favour of reduced public sector involvement.

As late as the 1960s, there were few multinational companies in the seed industry. A wave of acquisitions occurred in the 1970s as seed companies were bought up by transnational corporations, mainly food trading and petrochemical companies. Food traders, seeking to open up new export markets in the era of US 'food power', sought to extend their activities upstream.

The development of high-input seeds by international agricultural research centres had also led (agro-) chemical companies to seek new markets in the developing world, so these companies sought to market seeds through the same channels. With PBR legislation in place in many developed countries, seeds were no longer a weak link in the input supply industry. UPOV, by creating a degree of harmonisation in PBR legislation, fostered the emergence of a global seed industry, whilst the horizontal integration of agricultural input supply has deepened the inter-relatedness of inputs over time.

The Challenge to the BCM Paradigm

The BCM mode has come under, and continues to operate in the face of, considerable pressures for change. These are due to its:

- Ecological impact;
- Impact on food quality and health (in farming, and in consumption);
- Impact on rural communities;
- Being supported by state policies, and high levels of farm support, which effectively exacerbate the problems mentioned above.

It is beyond the scope of this book to address each of these in detail. Suffice to say that, in the words of Almas and Nygard, the BCM mode has produced 'some of its own executioners'.

Aims to reform agricultural and farm support policies seek, increasingly, to re-direct support towards environmentally sound practices. These are generally believed to imply reduced use of agrochemicals, and also seeds. To the extent that it continues to be allowed, seed-saving becomes more economically attractive in times of low output prices. In the spirit of challenges to the BCM mode, it is to the issue of seeds and germplasm that we now turn.

Sowing the Seeds of Discontent

As the Rockefeller and Ford Foundations expanded their efforts in international agricultural research, they began to seek public funds to support their work. Following a meeting in Belaggio in April 1969 organised by the Rockefeller Foundation, 15 governments attended the first meeting of the Consultative Group for International Agricultural Research (CGIAR). The World Bank would provide a secretariat and administer finances. International agricultural research, the year after Norman Borlaug received a Nobel Prize for his work on dwarf wheats in Mexico, had come of age.

In 1972, the CGIAR's Technical Assistance Committee (TAC) convened a meeting in Beltsville, USA, to formulate an international strategy for genetic resources conservation. After much debate, the International Board for Plant Genetic Resources (IBPGR) was set up as part of the CGIAR network. Under the IBPGR's system, the majority of the world's genebank accessions, mostly from developing countries given to believe that genetic resources were the common heritage of humankind, were stored in genebanks in developed nations increasingly predisposed to the notion of IPR over germplasm.

In the late 1970s, developing countries were increasingly concerned by IBPGR statistics showing that of more than 1.9 million samples stored, 55 per cent were in developed countries and another 14 per cent were held in the Northern (donor) dominated CGIAR system. Collections were clearly biased towards crops of interest to the developed countries, the top cereal crops representing more than 75 per cent of accessions in the pre-1980 period. At a 1981 IBPGR/FAO/UNEP conference, Latin American countries pressurised the IBPGR, successfully, to increase collections of crops less prominent in international trade.

By the end of 1981, despite opposition from US and UK representatives, a Resolution tabled by the Mexican delegation had been passed at the FAO calling for the FAO Director General to draft elements of an international convention on plant genetic resources, and investigate the feasibility of establishing a new international gene bank. Two years later, at the FAO's biennial conference, Jose Ramon Lopez Portillo, son of the former Mexican President, forced another vote which led to an International Undertaking on Plant Genetic Resources (IUPGR) and the creation of an International Commission on Plant Genetic Resources (ICPGR). The aim of these moves was transfer of control over genetic resources from the developed countries, and IBPGR in particular, to the United Nations.

Under the IUPGR, the notion of free exchange was to be respected, and it was not just landraces that were to be 'available without restriction', but also 'special genetic stocks (including elite and current breeders' lines and mutants)'. This angered the American Seed Trade Association (ASTA), who were represented in the US FAO delegation. They charged that the IUPGR struck:

> at the heart of free enterprise and intellectual property rights ... The definition includes unimproved and obsolete varieties, land races, wild and weedy species, all of which the seed industry believes appropriate to be preserved and freely exchanged. However, it also includes improved elite varieties and breeding lines within the definition of plant genetic resources ... This puts the Undertaking in direct conflict with the rights of holders of private property ... The anti-private business bias of the Undertaking is clear.

The IUPGR also proposed establishing a network of base collections under the jurisdiction of the FAO. Yet the Undertaking was a mild and voluntary agreement rather than a legally binding convention.

The Birth of Biotechnology

In the second half of the twentieth century, enormous strides have been made in the life sciences, particularly in the discipline which has come to be known as molecular biology. As a result of this growing body of knowledge, new commercial opportunities appeared on the horizon based on the use of tools developed through new discoveries in this field. The idea that plants could be made resistant to pesticides was no new idea. Wiebe and Hayes discussed it decades ago with regard to the reaction of barley varieties to the application of DDT. Yet work in biotechnology brought such a strategy closer to hand, raising the possibility of breeding plants designed to tolerate applications of proprietary chemicals.

The use of plasmids, in 1973 by Cohen and Boyer, to mediate gene transfer made possible a new alchemy. In the immediate aftermath, biotechnologists in the US showed awareness of public unease regarding this new technology by proposing a moratorium on certain types of research. Since 1977, guidelines laid down by the National Institutes of Health (NIH) have been progressively relaxed. The desire to regulate the industry has dwindled as authorities were persuaded of the commercial significance of the new technologies.

Reduced regulation of the biotechnology industry began to be perceived, and not just in the United States, as a means through which a country could maintain or improve its position in emerging bioindustries. Field reported that 'industrial competitiveness appears to represent the central and overriding concern of national strategies'. From a different perspective, the United Nations Centre on Transnational Corporations opined that comparative advantage and the international division of labour were increasingly being shaped by technological prowess. Increasingly, regulation shied away from determining which technologies should be allowed for use, and the imperative of allowing new technologies to develop began to shape which regulations were considered acceptable.

The emergence of biotechniques for technology generation makes it possible to circumvent the constraints imposed upon genetic recombination by species incompatability. I argue below that formal agricultural research is undergoing a transition from the BCM mode to a biotechniques-mechanisation-legislative (BML) mode. This is not to imply that traditional plant breeding and chemical inputs are about to disappear from view, either now or in the near future.

The role of breeders, where they do not disappear altogether, is likely to undergo a change such that their work complements that of the biotechnologists, whilst the fact that to date, herbicide tolerance in crops is the most widely tested trait to date testifies to the likelihood of continued use of chemicals into the future. Nevertheless, the genetic determinants of interactions between the plant and various chemical and biological inputs are likely to become the focus of innovation in crop (and livestock) agriculture. The pivotal institutional innovation in enabling such a strategy to become privately profitable is IPR legislation.

Patenting Genetic Materials

In 1976, the first of the new biotechnology companies, Genentech, was formed by Herbert Boyer and venture capitalist Robert Swanson. In 1980, Genentech placed a share offering on the New York Stock Exchange, the prices of which shot up from $35 to $89 per share in twenty minutes, a record rate of increase. This was due to the fact that three months earlier, General Electric had successfully challenged an earlier decision by the US Patent and Trademark Office (PTO) which had ruled that an oil-degrading micro-organism developed by their scientist, Ananda Chakrabarty, was not patentable subject matter. The new ruling held that whether or not an invention was alive or dead was irrelevant to patent law.

In the PTO's ruling on Chakrabarty, the legal principle of 'pre-emption' disqualified materials protectable under the PPA or the PVPA from patent protection. But this ruling was also overturned in the 1985 *Ex parte Hibberd* case, in which Hibberd was granted patents on the tissue culture, seed, and whole plant of a corn line selected from tissue culture. Breeders could now choose the form of protection most suitable to them, including utility patents. The gene was being commodified.

Anxious to preserve its lead in the biotechnological race, the United States has moved fastest in bringing institutions into line with industry's desires. Employing both bilateral and multilateral channels, it has sought to harmonise standards across nations in line with its own structures, thus opening the way for global marketing of proprietary products of biotechniques. In November of 1982, at a ministerial level meeting held at the GATT at US insistence, the US proposed that the new round debate issues never before considered in earlier GATT rounds.

One such issue was trade in counterfeit goods, such as 'fake' Rolex watches, but the scope of this particular area was widened at the behest of the US and others to include the issue of IPR. Raghavan notes that this was 'thanks mainly to the negligence of the disorganised Third World countries, most of whom thought that it did not affect them'. This would not have been so critical had it not been for the fact that what many countries saw as a preparatory discussion was subsequently proposed by the US and others as the agenda for a new round. The inclusion of many new issues was given justification through addition of the prefix 'trade-related'.

Many countries hoped that by stifling their objections to the inclusion of new issues such as 'services' and IPR, they would be rewarded with concessions on 'old' issues, such as tariffs on tropical products (and escalating tariffs on processed products thereof), textiles, and a continuation of benefits under the Generalised System of Prefences (GSP). A compromise text was agreed at Punta del Este in Uruguay at the end of 1986 which included IPR. Even then, developing countries refused to negotiate on the subject before the mid-term review in Montreal in December 1988, where agreement was reached concerning the negotiating agenda.

The GATT agreement was finally signed at Marrakesh in April 1994 despite the fact that market access negotiations had not been verified, and with many developing country negotiators complaining that they had seen the texts only weeks before. Ratification in many countries was rushed through with little debate, and on 1 January 1995, the new World Trade Organisation (WTO) came into existence. This would co-exist with the GATT until the end of 1995. The final text of the TRIPs (Trade Related Intellectual Property Rights) agreement establishes new multilateral rules on IPR based on uniform minimum standards for their protection and enforcement, including their availability, use and scope.

As regards plant materials, the treaty does allow for exemptions on grounds of perceived environmental or public order impacts, yet at the same time, the treaty states that plant varieties shall be protected by patents 'or by an effective *sui generis* system or by any combination thereof'. Although developing countries and least developed countries are allowed, respectively, five and ten years to implement the agreement, the '*sui generis*' clause is to be reviewed four years after the date of entry into force of the WTO agreement.

The TRIPs agreement offers little encouragement to communities that might seek to protect innovations which are the property of, as it were, the collective. The agreement recognises IPR as private rights, and also requires products to be 'capable of industrial application'. There is no mention of communities and their rights.

Before the GATT negotiations even began, the US had made its intentions in respect of IPR abundantly clear through applying pressure bilaterally. Mexico was targeted as early as the 1970s, but little progress was made until, in the mid-1980s, the US began to link the issue of IPR reform to expansion of GSP concessions. By the end of 1986, Mexico had adopted a revised Patent and Trademark Law, though as a result of the efforts of domestic lobbying, this was deemed inadequate by the US administration.

In 1984, the US Trade and Tariff Act had been revised, S.301 of Title III of which invested the prevailing Administration with coercive powers aimed at righting 'unfair' trading practices. In 1985, cases against Brazil and South Korea were initiated, the former concerning, *inter alia*, copyright issues regarding software, the latter concerning failure to protect intellectual property. The same issue led to talks with China, whilst India has also come under pressure to reform its IPR legislation in the past.

In January 1987, Mexico was informed of President Reagan's intention to withdraw $200 million of GSP benefits unless the perceived inadequacies of its new legislation were corrected. In 1988, the Omnibus Trade and Competitiveness Act was passed in the US. This included sections which came to be known as Super 301 and Special 301 respectively, retrospectively strengthening the coercive powers vested in the administration by the Trade and Tariff Act's 1974 revision. Under Special 301, a procedure was set up whereby the US Trade Representative could identify and initiate proceedings against countries considered to be offering inadequate IPR protection.

Within this list of countries, a Priority Watch List of countries was to be specified annually, and Mexico was on that first list in May 1989. When President Salinas de Gortari began his programme of liberalisation, and plans for a North American Free Trade Agreement were materialising, Mexico's stance on the IPR issue altered quite radically, and in 1990, when Mexico introduced a proposal for a TRIPs agreement at the GATT, it slipped off the 301 lists. Mexico's patent law was revised in June 1991 to explicitly allow for the patenting of plant varieties. It specifically addresses innovations likely to arise from the deployment of biotechnology.

As Fowler has made clear, GATT and the growing concerns of the US over intellectual property issues generally, enabled transnational corporations involved in agricultural biotechnology to have their concerns *vis-à-vis* the patenting of life addressed in new fora. In the case of the GATT, it became possible for the issue to be bundled up not only with concerns over patents and trademarks as they related to mechanical innovations, but also, since this was a take-all-of-it-or-leave-all-of-it package, with fourteen other areas with which the Uruguay Round was concerned.

Significantly, Watkins notes, 'The major actors in [the TRIPs] exercise have been the US-based Intellectual Property Coalition – a grouping of 13 major companies, including IBM, DuPont and General Motors – *and European agro-chemical giants such as Unilever, Hoechst and Ciba Geigy*.' Thus, IPR was being simultaneously harmonised and extended across the globe.

Upov

Paralleling the moves to enhance intellectual property protection under the auspices of the GATT were moves on the part of UPOV to bring the Convention into line with developments elsewhere, and particularly with respect to biotechnology. PBRs' research exemption made them inadequate for protecting biotechnically engineered plants since they offered protection at the level of the whole plant when what was required was protection at the level of the gene. But by 1987, it was clear that UPOV would be strengthened. According to Fowler *et al.*, UPOV's members had been divided between small seed houses and the integrated genetics supply industry, the former fearing gene patenting, the latter favouring new initiatives in this respect. UPOV was revised in March 1991.

Note that the right of farmers to save seed from one harvest for planting in the next, what the American Seed Trade Association had referred to as the 'farmers' right' in hearings on the PVPA, had become known as the 'farmers' privilege' and was no longer secure. Section 15.2 of the new Convention allows, as an optional exception, seed saving 'subject to the safeguarding of the legitimate interests of the breeder', implying that royalties should be paid to breeders where seed is saved.

On the other hand, there is a compulsory exemption for breeding other varieties. However, the interests of the breeder are, in general, strengthened since protection applies to 'Essentially derived and certain other varieties' as defined under Article 14 (5) (b) and (c). Lesser expresses concern that, since the definition is unclear, this will lead to quasi-IPR being granted to a breeder over thousands of attributes of a variety which he/she did nothing to create. The 'essentially derived' clause would appear to apply to genetic insertion, giving the owner of PBR the right to demand royalties from innovations based on insertion of one or two genes into a plant over which the right is held.

Increasingly, developed countries are bringing their PBR legislation into line with the 1991 UPOV Convention. Furthermore, although the TRIPs '*sui generis*' clause appears to offer room for manoeuvre in designing IPR for plant varieties, most believe that this translates into UPOV type standards and nothing less. Increasingly today, companies can choose which combination of protection they prefer, although European patent law currently forbids patenting of plant varieties.

LOCKING IN TO UNIFORMITY

Reflecting the above-mentioned events, the Crucible Group reported: 'Those who reviewed patent law a few decades ago may not recognize it today'. Probably no formal agricultural research organisation in the world has not at least cogitated upon the changes considered above, not all of whose are implications are, as yet, clear. It is important to understand at least some of these, and to contemplate the relevance of the changes to the existing BCM mode of agriculture.

Biotechniques and environmental critiques of agriculture are reported to be bringing about a reshaping of the technological development of agriculture. This is jumping the gun slightly. There are still unresolved questions concerning, in particular, consumer acceptance of genetically engineered products, which are already having an impact on the industry's development.

However, Sharp may be right to talk of the laying of a new set of 'ground rules', making it 'inconceivable that those developing new drugs, new herbicides or pesticides, or new plant species, should not, somewhere *en route*, make use of gene cloning and sequencing techniques.' The implications would be that research which did not require such techniques might fall by the wayside.

A number of authors have commented on the paradigm-like shift that biotechnologies could achieve. Much of this discussion considers the issue at the macroeconomic level, and takes the view that it will not be biotechnology alone that leads to a new mode of accumulation, but biotechnology, the development of new materials, and information technologies working synergistically to form a new techno-economic paradigm.

There is no doubting that there could well be some revolutionary changes about to occur in the way in which the agro-food system functions. Most interesting of all are potential developments in the food processing industry, where some authors have speculated as to the possible emergence of a 'generic biomass inputs sector' as a result of technologies which allow biological materials to be fractionated into component parts for the final manufacture of food products.

The implications for commodity markets as they are currently understood could be far-reaching. Other potentially revolutionary techniques relate to so-called novel products, which will affect the ways in which agriculture interacts with other sectors of the economy.

Yet, whilst certain techniques used to create new products are certainly emerging, there appears to be substantial continuity with the past with respect to:

- Increasing horizontal integration across agricultural inputs – breeding for responsiveness to inputs and to facilitate harvesting will give way to herbicide tolerant varieties.
- Deepening of vertical integration – breeding has facilitated mechanised harvesting and handling of the final product. Biotechniques are increasingly geared towards downstream aspects of food production, representing higher value-added, and greater opportunities for profit, in upstream sectors of the

chain of value in food. Lamola speaks of end-use tailored, or identity-preserved varieties.

- The actors involved are, in many cases, one and the same as those who prospered through the BCM paradigm (erstwhile agrochemical and seed companies).
- Emerging products take their cue from their supposed ability either to replace, or alter the functioning of, elements of the BCM paradigm which have been heavily criticised in the past.
- A continuing lack of emphasis, in private sector breeding, on pest resistance – although biotechniques provide tools for reducing pesticide use, current trends seem likely to increase, rather than reduce their use. Where resistance breeding is undertaken, it is of the gene-for-gene, vertical resistance type.

In many respects, therefore, the goals remain rather similar to those in the BCM paradigm. In particular, the attractions of the new techniques are seen principally in terms of the increased control that can be exerted over the transformation of organisms through recombinant DNA techniques. Indeed, Richards speaks of biotechniques as heralding a 'second designer phase' for agriculture.

Whereas the Green Revolution focused on ideotypes for monocropping in controlled *physical* environments, the second phase seeks to shape 'econotypes' to meet the need of future *economic* environments. As with the BCM mode, an emphasis on control within the labouratory has tended to obscure and marginalise the significance ecological issues concerning the functioning of biotechnically engineered products in the field.

The most obvious break with the past is the ability of biotechniques to extend the genepool available to breeders and biotechnicians beyond the primary and secondary genepoools into the hitherto unexplored (because of species incompatibility) tertiary genepool. Less immediately obvious, are the changes which have already been wrought by biotechniques on our perception of the nature of life itself, and the significance of these for our perception of the nature of food and agriculture in the longer-term.

Paradoxically, therefore, we are witnessing changes which are simultaneously profound, and incremental. The mode of agricultural research is changing through use of powerful new techniques, but its roots remain in the BCM mode. This is to be expected if one accepts that agriculture had become locked in to the BCM way of doing things. As Teece points out, one aspect of the locking in process is that firms tend to do best what they have done in the past.

If the emergence of biotechnology constituted a radically new paradigm, learning advantages accumulated over time by established FAROs would have lost much of their significance. However, the fact that biotechnology is very much a process technology has meant that much of the significance of learning, particularly in the downstream operations of private multinational corporations, has been retained. Furthermore, as I have suggested above, the techniques are deployed in pursuit of a familiar goal, that of the genetically uniform ideal plant.

Changing Modes

The transition that is occurring can be understood through appealing to the framework developed by Freeman and Perez. As noted above, the BCM mode has come under fire for a variety of reasons, principally those associated with food quality and the environment. If the limits to the expansion of this mode had not yet been reached, such expansion was clearly under threat. The world market for agrochemicals saw three years of decline in the years 1991–93 before recovering somewhat in 1994 and 1995.

Farm support schemes, in the European Union and elsewhere, have begun to shift away from price support, which led to elevated levels of use (relative to that which would prevail with prices at world market levels) of agrochemical inputs, and towards conservation, often rewarding farmers for using fewer inputs.

For Freeman and Perez, the transition from one techno-economic paradigm to another, brought on by the onset of recessionary trends, is characterised by:

The increasing degree of mismatch between the techno-economic sub-system and the old socio-institutional framework. It shows the need for a full-scale re-accommodation of social behaviour and institutions to suit the requirements and the potential of a shift which has already taken place to a considerable extent in some areas of the techno-economic sphere. This re-accommodation occurs as a result of a process of political search, experimentation and adaptation.

Once the socio-institutional framework matches the techno-economic sub-system, investment moves forward and growth is restored. For the BML mode to flourish, its techno-economic sub-system requires an appropriately matching institutional framework, including:

- a sympathetic regulatory framework for undertaking relevant research, including risk assessment methodologies as applied to the release of the products of gene technologies, and food safety legislation regarding genetically engineered food;
- political, social and environmental acceptance of the technologies and their end-products, reflecting confidence in the regulatory framework; and
- appropriate IPR protection, the importance of which is confirmed by, amongst others, Thelwall and Clucas, Caulder, Lamola, and Duffey.

A lack of institutional change will delay any upswing, and indeed, resistance from consumers concerning issues of health and environmental risk has been strong. Consequently, products have been slow to reach the market. Yet, for reasons elabourated below, it is the issue of IPR legislation which has greatest bearing on the issue of biological diversity in use in agriculture, and thus, the environmental risks posed by new biotechnologies in terms of vulnerability and the continued use of pesticides.

Those in the vanguard of the BML mode have sought to project it as environmentally friendly. In doing this, they have stressed the biological, *ergo* natural, characteristics of the work they are undertaking. The semantics involved have been illuminating, at one and the same time suggesting radical new possibilities (the economic and environmental attractions) and on the other, in an attempt to downplay the risks associated with the products of biotechniques (and the need for regulation), suggesting continuity with the past. Critics of biotechniques turn the matter around completely.

Whilst not disputing the fact that there is money to be made, they argue that the new possibilities should be reflected in the need for new forms of regulation, whilst continuity is likely to be reflected in the continuation of environmental problems. Their criticisms relate mainly to:

- The uncertainties in *ex ante* risk assessment associated with release of genetically engineered organisms into the environment, not least the difficulties in extrapolating from small-scale trials to large-scale field use, and problems associated with trade in the products concerned.
- The nature of individual products and their possible environmental and health consequences
- The environmental consequences of possibly increased uniformity.
- The political economy of the research being undertaken (who is it done by, and for?).
- The impact in terms of research not undertaken (a point well made by Rachel Carson in 'Silent Spring'.

These issues are not unrelated. The nature of the organisation funding research will determine the degree to which a notional social welfare function is reflected in their activities. Private organisations need not be concerned with social welfare, or only insofar as it affects profits.

The key to deepening private sector involvement in agro-food biotechnical research has been the extension of IPRs to living organisms. Governments have welcomed private sector participation in research, and have tended either to move the focus of their research away from near-market, and towards more basic, research, and/or to seek to take advantage of the patent system themselves to make financial gains from ongoing research.

Future for Diversity

As mentioned above, one of the criticisms levelled at those who believe that biotechniques herald a new 'sustainable agriculture' is that existing problems of uniformity will be exacerbated. Can the transition from BCM to BML mode re-introduce diversity into a system based on uniformity? From a purely technical view, biotechniques' capacity to draw upon genetic material from the tertiary genepool would suggest that additional genetic variation might be introduced. Thus, Bassett argues that new varieties 'will simply coexist with the old varieties: diversity will have been increased, not decreased'.

But this approach has two major shortcomings. Firstly, the basic research and the application are inextricably linked, so much so that possible applications are driving the direction of basic research. Thus, a growing proportion of public sector research is supporting commercially oriented research. Secondly, what actually happens is a subset of what could happen, as the earlier case-studies have argued. Kloppenburg bluntly states: 'the baby of biotechnology is not so easily separated from the corporate bath water,' which is exactly why, as Lesser points out, trends towards uniformity predate the existence of PBR and patents.

Strengthening IPR as applied to living material has encouraged private industry to engage in biotechnical research at levels above those that would have existed in their absence. Furthermore, because Governments now see biotechnical prowess as important for maintaining competitiveness, public research is beginning to resemble privately sponsored research, either through its increasingly subservient position to private industry, or through more overt aims to generate revenue from patentable research outputs. Erstwhile President of Harvard University, Derek Bok, has expressed concern that Universities will increasingly 'differ from corporations only because there are no shareholders and no dividends'.

An element of historical contingency is at work here, since many governments are cutting state spending on education and research where it is perceived to be of low market value. To the extent that environmental issues are intimately related to issues of social welfare, and because many environmental costs are not captured in market transactions, one assumes that private organisations are, notwithstanding their own public relations, less likely to integrate environmental issues into their research programmes.

Indeed, one survey, aimed at eliciting the ranking of breeding companies' priorities, placed the environment at the bottom of seven criteria. However, the same is increasingly true of publicly funded research. Institutional changes are making the public / private distinction irrelevant in predicting the social welfare goals which will be pursued by one or other form.

This is one reason why, notwithstanding some of the claims made for the efficacy of biotechnologies (most of which, incidentally, are made with the BCM technologies as the implied baseline), IPR will if not increase, then maintain, the vulnerability of agriculture in the field. Other reasons include the following:

Distinctness of Varieties

Already, anecdotal evidence suggests that seed companies rely on a few elite cultivars in their research programmes and new varieties are developed through minor modifications to these. Very little hard evidence is available, but it is clear that the number of varieties available (i) does not reflect genetic diversity, and (ii) masks the concentration in varietal use out in the field.

The future strategy for breeders will be structured by the IPR legislation in place. The combination of UPOV 1991 and patents ensures that the work of the breeder, and the value of an identified gene, is both recognised. The 'essentially derived' clause was introduced to deal with the problem of genetic distancing. Biotechniques make it possible to reduce the genetic distancing required to discriminate between varieties, in which case, the genetic variation existing in the field would become even further divorced from consideration of the number of varieties grown.

For example, since distinctness could now be measured at the level of the gene, a superfluous (in agronomic terms) gene could be spliced into a variety thus making it, potentially, distinct. As Smith points out, the practice of reverse engineering of varieties is increasingly common and would make the aforementioned practice more likely, rendering the granting of PBR meaningless. The 'essentially derived' clause, though of theoretical value, raises important questions of definition. For both Espenhain and Smith, who, in his excellent account, notes that numerous controversies in this regard are in no one's interest, case law will provide the answers.

Because UPOV 1991 extends the breeder's right to the commercialisation of essentially derived varieties (the principle of dependence), companies using genetic transformation techniques will either work with their own PBR-protected varieties, or license genes of interest to other companies for incorporation into their varieties. Strategic alliances between those specialising in biotechniques and those with greater specialism in traditional breeding seem likely.

Hence, Pioneer markets both soyabeans containing Monsanto's Roundup Ready gene (tolerant of Monsanto's glyphosate herbicide) and soyabeans containing the DuPont-owned sulfonylurea tolerant gene. DEKALB also has cross-licensing agreements with both Monsanto and DuPont. Those licensing technologies will seek to gain from the technology premium which biotechnically developed products aim to attract.

To a significant degree, varietal make-up will remain as before, but with genes spliced into a particular variety's background, and with plants themselves being made more uniform. However, since industrial structure will be affected by the evolving IPR framework, as well as the techniques themselves, and developments in individual sectors, there may be implications for the diversity of what is offered to farmers. This is considered in what follows.

Research Concentration

Patent enforcement prevents companies from carrying out research on a patented genetic sequence or process without paying royalties to the patent holder, whilst essentially derived varieties are subject to PBR under UPOV 1991. The strategic importance of patents for any company depends to a considerable degree on the extent of exclusion implied by the text of the patent. In this context, the current trend towards granting broad patents to companies is worrying indeed.

For example, Agracetus, a subsidiary of W. R. Grace (and recently taken over by Monsanto), was granted patent rights in the US over any genetically manipulated variety of cotton, and by the European Patent Office over all genetically transformed soybeans (with those for rice, groundnut and maize pending). Agrigenetics' patent on high oleic acid sunflowers effectively stopped all such work in this area outside the company. The prevalence of 'driftnet patenting' is at odds with the view held by many that patents encourage innovation. It raises the possibility that the seed industry for any one crop may ultimately become dominated, or at least hostage to, one commercial enterprise.

Patent enforcement is an important tool in building corporate empires and eliminating competition. Monsanto, which has staked much on its quest to become the 'Microsoft of engineered foods', has acquired companies, and stakes in others, reflecting its belief that patents will be a key source of competitive advantage in coming years. Pioneer's moves to patent its in-bred lines on grounds that this would prevent other companies carrying out research on them illustrates that even the 'biological patent' which hybrid corn varieties are endowed with is being superseded by strengthened IPR legislation.

As patents proliferate, it will become increasingly difficult for any enterprise to conduct research in the full knowledge that it is legitimate, especially since burden-of-proof legislation makes it incumbent upon those accused of patent violation to prove that they are innocent. The possibility arises in which a company carrying out research happens, by accident, to be working with a variety in which a patented gene sequence exists. Such a company would be unwittingly breaking the law, and would be expected to provide proof of its innocence.

There is a suggestion that IPR-related concerns are driving strategic alliances in the industry. Since, increasingly, more than one form of IPR will be involved in developing a given variety, such alliances reduce the likelihood of any one IPR-holder blocking development of the product concerned. IPR-based restructuring can be expected to produce 'many casualties, some survivors, and a few successes'. The growing significance of IPR, appears to be leading to greater concentration in breeding effort, which is unlikely to promote diversity in agriculture.

Loss of Farmers' Privilege

UPOV 1991 appears to deny that farmers might also be breeders. Industry estimates suggest farmer-saved seed accounts for between one and two thirds of all seed planted in the world, though the proportion for developing countries is believed to be of the order 85 per cent. Indeed, in developing countries, the exchange of seed from farmer to farmer is probably the main avenue for diffusion of new seed varieties.

Erstwhile Director General of GATT, Peter Sutherland, has suggested that such informal practices are 'generally not of interest to the owners of protected varieties', yet already in the US, court actions have been taken against farmers involved in such activity, and it will not have been lost on IPR owners that interfarm sales constitute 62 per cent of all seed purchases in India where Sutherland made his speech.

In the US, companies such as Monsanto require growers of their Roundup Ready Soybeans to be licensed to grow the material, though other companies, including AgrEvo and DuPont, do not require licensing since, in the words of one commentator, 'they seem to feel that the additional chemical usage that's tied in with [DuPont's] STS beans is enough'. Monsanto states that 'if necessary, the terms of the [Roundup Ready] contract will be enforced under the P.V.P.A., US Patent Law and general contract law'. One study estimates that within a few years, 40 per cent of all US farmers will be contract growers, or renters of germplasm from the same companies to whom they sell their product.

Farmers will also need to be alive to the possibility, especially when growing out breeding crops (those which cross-pollinate) of falling foul of patent legislation. It is not clear, given reverse burden of proof, how the law would interpret a situation where a farmer grew a variety which through cross-pollination contained a patented sequence. Potentially, the onus will be placed upon farmers to ensure no such cross-pollination takes place. Furthermore, the possibilities for farmers to experiment with varieties covered by UPOV 1991 seem limited.

As with the case of 'essentially derived' varieties, it seems likely that case law will determine what is and is not allowed, but in the meantime, farmer experimentation with new varieties, which can create new races of out-breeding crops, may be a risky enterprise. This may have implications for diversity. More recently, the patent awarded to Delta and Pine Land Company on so-called 'terminator technology' (which prevents seeds from germinating in the next generation) provides a technological means through which to prevent farmer seed-saving.

Wide Use of Agronomic Genes

It has been stated that the introduction of novel genes will increase genetic diversity in agriculture. However, if the same gene is licensed to several companies for use in a large proportion of varieties in use, and in different crops, the gene becomes a component of uniformity. The possibility arises of the occurrence of a southern corn leaf blight on a more global scale. In China, 15 million hectares are planted to hybrid rice, each plant possessing, as with US corn varieties in 1970, a common gene for cytoplasmic male sterility.

Genes from the bacterium, *Bacillus thuringiensis,* which produce a protein that is lethal to some insects upon ingestion, are of great interest to corporations involved in biopesticides. Yet there are already concerns for insects' resistance to a number of strains of the bacterium.

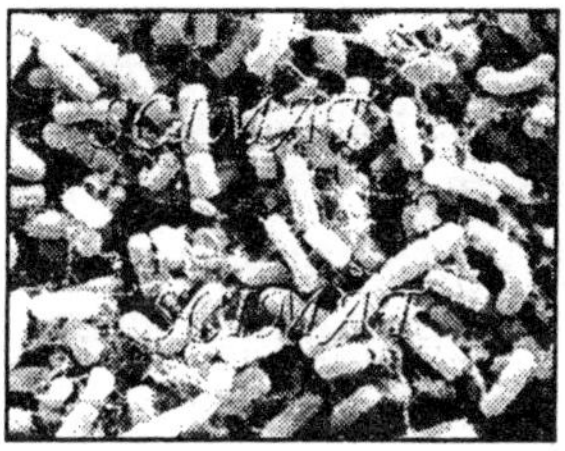

Fig. Bacillus thuringiensis

Herbicide tolerant genes could quite conceivably be transferred into vast areas of crop land, potentially increasing the vulnerability of crops globally, and leading to heightened problems with herbicide resistant weeds. Indeed, the relay race mentality of the BCM mode is accepted as a matter of course: 'in 50 years, biotechnologists will almost certainly still be developing new batteries of pest- and disease-resistant genes'.

The aim to engineer plants with tolerance to herbicides is a goal of companies that have integrated crop protection and seed production. In this way, purchasers of seed would be locked-in to the purchase of proprietary chemicals.

This strategy reflects the fact that the costs of developing new agrochemicals are increasing owing to costly approvals processes. The costs of engineering the seed to lengthen the effective life of a given chemical compound are less, whilst it is also possible to extend the patent life of chemicals coming 'off patent' by specifying use of the proprietary form of a generic compound. The industry claims to have shifted emphasis to compounds of lower toxicity, although much attention has focused on glyphosate, which was listed by the US National Academy of Sciences as a potential carcinogen in 1987.

Ecological Interactions

One of the major fears of environmentalists is that a gene which has been transferred to a variety to enhance its competitive performance may be transferred sexually into wild relatives, especially where maize, potatoes, rice, chickpeas and common beans are concerned.

In these crops a wild-weed-crop complex is observed in which there is continual gene flow between wild and cultivated forms. More speculatively (the processes are poorly understood), horizontal gene flow mediated by micro-organisms may occur. In either case, the transfer of one gene may be sufficient for a plant to become invasive.

Lack of Research on Diversity

Quite apart from the tendencies remarked upon above which might exacerbate trends towards uniformity, the simple fact remains that little if any research associated with the use of biotechniques is being undertaken to encourage the use of diversity in the field. Indeed, biotechniques make it possible to clone plants and seeds so eliminating what residual variability there may have been in a crop bred using traditional methods.

Existing Seed Marketing Legislation

Perhaps most importantly, and what may in time become the most concrete expression of the way in which biotechniques affect the offerings of the seed industry to farmers in the fact that over time, varieties which are not the product of genetic manipulation will slip off existing National Lists of seed varieties. Since those on the list are the only ones which can enter into commerce, slowly it will become impossible to purchase seeds which are not genetically engineered.

This will force farmers and consumers alike, irrespective of concerns regarding genetic engineering, to purchase genetically engineered seeds. To the extent that genetic engineering concentrates on the integration of specific sequences within existing elite cultivars, irrespective of the number of cultivars made available to farmers, the genetic diversity within farming may decline, and certainly seems unlikely to increase.

Possibly, IPR will not increase uniformity. One scenario would see patented genes integrated into the background of existing varieties, and no change other than those created by the introduced genes. But this scenario assumes an unlikely scenario in which changes wrought by IPR and biotechniques will leave the seed industry unchanged in other respects, a scenario which current trends suggest is unlikely.

Our inability to measure diversity, and the fact that we do not know where we stand today, makes it impossible to assess change on the basis of any reliable baseline. However, it seems reasonable to suggest that biotechniques are taking agriculture in different directions to those which would be implied by the alternative approaches.

If the BCM mode was environmentally damaging, and if there remain unanswered questions regarding the impact of the BML mode, why does this new mode appear to be gaining the support of most FAROs, public and private, especially when alternative paths exist?

ALTERNATIVE AGRICULTURE

In Mellon's words 'the hype surrounding biotechnology diverts our attention from those [pesticide free] solutions by focusing attention on technologically dazzling new products. By setting the proper goal, we will avoid the danger of spending millions trying to genetically engineer ten "better" pesticides, when for far less we could have taken our agriculture systems off the pesticide treadmill forever.'

In particular, the alternatives (re-)emerged with some force in the 1980s on the back of environmental concern seem likely to remain alternatives in terms of the resources devoted to them. The transition from BCM to BML mode must be understood in this context. This was not a transition that was inevitable. The BCM was under fire, and alternative approaches beyond biotechniques were available.

If IPR was essential to the BML mode, they may be decidedly unhelpful for those seeking to do research outside that mode. Sederoff and Meagher opine that IPR 'are having a dramatic negative effect on the progress of non-profit research'. The following sections consider debates concerning genetic resources which have taken place outside GATT and UPOV.

Recognition for farmers' rights

The aim to resist IPR strengthening has been closely related to attempts to encourage the use of diversity in agriculture. Whilst the agricultural biotechnology companies had their sights on institutional changes allowing for the granting of IPR over life through the GATT, the FAO was debating the issue of the rights of farmers over germplasm, especially from 1987 onward.

Farmers' Rights would be the counterbalance to the spread of PBR. In 1987, an International Fund for Plant Genetic Resources was set up, and was legally established in 1989. The Fund was designed for genetic conservation and utilisation work, and administered by the ICPGR. Mexico argued that donations to such a Fund should be mandatory in the same way as are royalties to a patent holder, but no agreement was forthcoming.

By 1989, developing country governments had let it be known through negotiations at the FAO that failure on the part of developed country governments to acknowledge the concept of Farmers' Rights would result in those countries being denied access to developing country genetic resources. Similarly, developed country patents would not be honoured.

In what was effectively an exercise in horse-trading, at the 1989 meeting, developed countries insisted on an additional Resolution modifying the International Undertaking such that it recognised PBR. 'Free accesses to germplasm explicitly would not mean 'free of charge'. In return, a Resolution on Farmers' Rights was passed recognising the rights of farmers in respect of their work conserving, improving, and making available genetic resources.

In the midst of an increasingly polarised debate, a notable event was the Keystone Dialogue held at the Keystone Centre in Colorado in 1988, at Madras in 1990, and Oslo in 1991. Major transnational corporations, NGOs, IBPGR, national genetic resource programmes, the academic community, and the Rockefeller Foundation reached the following consensus conclusions after the second meeting:

- IPR were credited with encouraging development of new varieties, but also with encouraging genetic uniformity and erosion;
- Attempts to include IPR for plants under the auspices of the GATT negotiations were criticised; and
- Recognition of Farmers' Rights, and commendation of the idea of a Fund such as that extant at the FAO as a means of providing a form of concrete recognition thereof.

Corporate delegates refused to sanction a compensation mechanism, merely a fund recognising Farmers' Rights, a position which some participants would have found unacceptable were it not for the fact that it was agreed the fund should be mandatory. Furthermore, rather than the figure of $150 million proposed by NGOs at the FAO, the consensus figure arrived at was $500 million.

In November 1991, another Resolution concerning genetic resources was passed at the FAO. This amendment to the Undertaking upheld 'that nations have sovereign rights over their plant genetic resources and that breeders' lines and farmers' breeding material should only be available at the discretion of their developers during the period of development'. Although Farmers' Rights were recognised, they were given no substance. The International Fund, legally established in 1989, failed to materialise. Only in India have attempts been made to give substance to the concept through taxation of seed industry income.

Farmers' Rights still amount to little more than a polite thank you to farmers who have conserved genetic diversity *in situ*. This was most clearly illustrated in discussions at the June 1996 Leipzig Conference where the issue of Farmers' Rights showed that developed country donors were reluctant to support *in situ* conservation, partly, one suspects, because of issues related to sovereignty in respect of genetic resources (see next section).

The Convention on Biological Diversity

National sovereignty over genetic resources was a feature of the Convention on Biological Diversity (CBD), which entered into force as a legally binding international treaty at the end of 1993. The CBD is a framework convention whose objectives, stated in Article 1, are:

the conservation of biological diversity, the sustainable use of its components and *the fair and equitable sharing of the benefits arising out of the utilization of genetic resources, including by appropriate access to genetic resources and by appropriate transfer of relevant technologies*, taking into account all rights over those resources and to technologies, and by appropriate funding.

Article 3 of the CBD lays down the principle that:

States have, in accordance with the Charter of the United Nations and the principle of international law, the *sovereign right to exploit their own resources* pursuant to their own environmental policies, and the responsibility to ensure that activities within their own jurisdiction or control do not cause damage to the environment of other states or of areas beyond the limits of national jurisdiction.

This effectively confirms that the free exchange principle is something of the past.

For corporations, sovereignty appeared to cede too much control to governments over genetic resources which were of increasing value to biotechnology companies. For non-government organisations, this debate seemed to miss the point that it was not states who were really responsible for maintaining biological diversity within their borders, but local communities. The CBD offers little for local communities, and does not explicitly recognise Farmers' Rights, though the role of indigenous communities in conserving biodiversity is recognised in the preamble.

Articles 12, 17, 18 and 19 each refer to aspects of the biotechnology debate, but Article 16 of the CBD, dealing with transfer of technology, was the most heavily negotiated. Essentially, the debate centred around the fact that developing countries would most likely be providing raw materials for a biotechnology industry seeking to patent innovations. Article 16.2, suggests that:

In the case of technology subject to patents and other intellectual property rights, such access and transfer shall be provided on terms which recognise and are consistent with the adequate and effective protection of intellectual property rights. The application of this paragraph shall be consistent with paragraphs 3, 4 and 5 below.

This Article was essentially a compromise meant to defuse the situation as regards the way in which technology transfer should account for IPR. These latter paragraphs make provision for the transfer of technology to developing countries on 'mutually agreed terms'. Article 16.5 suggests that the Contracting parties:

recognizing that patents and other intellectual property rights may have an influence on then implementation of this Convention, shall cooperate in this regard subject to national legislation and international law in order to ensure that such rights are supportive of and do not run counter to its objectives.

The CBD's equivocation on IPR issues was perhaps the main reason why George Bush's US delegation felt unable to support the CBD in Rio. Ever since the late 1980s, when methods of screening increased in sophistication and fell in cost, plant research has acquired new significance for pharmaceutical companies. Thus, head of the US delegation, William Reilly, stated: 'We have negotiated in the Uruguay Round of GATT to try to protect intellectual property rights. We're not about to trade away here in an environmental treaty what we worked so hard to protect there'.

The CBD does not apply retrospectively. Thus, the legal status of *ex-situ* collections of resources donated by developing countries but housed in the developed world was left unclear. In 1994, the World Bank appears to have sought to prevent the CGIAR's *ex situ* collections from falling under intergovernmental control by taking control of these itself in exchange for new funds for the CGIAR.

However, 112 governments unanimously, and ultimately successfully, called for establishment of intergovernmental control over the CGIAR *ex situ* collections. The status of other collections held in developed country genebanks was discussed at the FAO's Leipzig Conference, but the outcome gave little encouragement to those countries which donated germplasm in the first place, as exemplified by the attempts of the company PHYTERA to gain access to genetic resources collected in developed country botanical gardens.

Indigenous Peoples

As so often in the past, indigenous peoples have been forgotten in the bulk of negotiations which affect a resource maintained largely by them. Their concerns span both the prime focus of the CBD and the issue of Farmers' Rights, as well as the TRIPs negotiations. Increasingly, the contributions made by indigenous peoples in terms not just of germplasm, but also knowledge regarding its use, are recognised.

Yet this recognition has led to little concrete action to protect their interests. In response principally to the heated debate generated by issues related to the patenting of indigenous people's cell lines, the Indigenous Peoples Biodiversity Network was formed with the objective of safeguarding their interests with respect to biodiversity and their knowledge thereof. Out of the debate concerning Farmers' Rights and IPR as they affect indigenous peoples has come awareness that current formal systems do not adequately recognise indigenous knowledge systems, in which knowledge is often held at the level of the community.

More and more companies are screening plant materials for useful products, yet it is estimated that the chances of finding useful products are at least doubled if indigenous, or folk knowledge is utilised. An authority on the issue of IPR and indigenous peoples, Darrell Posey, notes that IPR pose seven problems for indigenous peoples:

- They do not grant rights to collective entities;
- They protect unique acts of discovery rather than transgenerational knowledge (from, for example, spirits, vision quests, or oral transmission) which tends to be public;
- They do not recognise non-western systems of ownership, access and tenure;
- They aim to promote commercialisation whereas the aims of indigenous peoples may be to prevent such activity;
- They recognise market values only and not spiritual, aesthetic, or cultural value;
- They are, as is clear from the above, intimately bound up with power relations;
- They are expensive to obtain and difficult to defend.

Posey goes on to cite a number of examples where indigenous peoples have displayed almost uniform hostility to what they perceive as an insidious trend in IPR legislation which seeks to deepen the exploitation of their resources and their knowledge. They perceive simple recognition of their contributions, as exemplified by the Farmers' Rights issue, as typically patronising in the face of a continued absence of legal mechanisms adapted to meet their needs and concerns.

RAFI explores a number of ways in which indigenous communities could find space within existing legislation to protect their innovations. They argue that:

> There is a strong case to be made that the uncompensated appropriation of farmers' varieties and medicinal plants constitutes real theft and that the parties responsible should be pursued under criminal law at the expense of national law enforcement agencies in the country where the theft occurs (the patenting country).

In the face of the forces mentioned above, it seems unlikely that much room for manoeuvre exists for those who would seek to place Farmers' Rights on the same level as IPR. Some countries have, however, been exploring the potential for exploiting the '*sui generis*' clause mentioned above in designing alternative IPR regimes which allow, for example, for communities, and not just individuals, to make IPR claims. As pointed out by Allen, much innovation is the product of collective rather than individual efforts. In India, the concept of Collective Intellectual Property Rights is gaining credibility, whilst the Andean Pact is committed to developing a regime on collective rights of indigenous peoples.

This chapter has sought to show, through examining evolving IPR regimes, that a new mode of development of agricultural technologies is emerging. Some, indeed most of its most vocal and powerful supporters are drawn from the leaders of the BCM mode. The commercial possibilities presented by biotechniques saw pressure to extend IPR schemes. Following the 1980s, a decade which saw recognition gained in an international forum for the concept of Farmers' Rights, the rights of farmers to save seed first underwent conversion to a 'privilege', and were then consigned to history.

The old BCM mode still prevails despite the attacks of environmentalists and the growing awareness of available alternatives. If the possibilities for alternatives to thrive alongside the BCM mode seem limited, they are likely to be more so as the BCM mode is superseded by BML techniques. Although the BML mode's supporters have often appeared as keen as environmentalists to see the back of the old BCM mode, their agenda is not an alternative based on bringing diversity back into the picture, but the ushering in of new techniques aimed at increasing the potential for achieving a genetically uniform ideal plant.

Major suppliers of agrochemicals condemn the technologies for which they themselves have been responsible as manifestly unsustainable. In answer to his own question whether such companies are 'Planetary patriots or sophisticated scoundrels?' Kloppenburg writes: 'Having been recognised as wolves, the industrial semioticians (and you thought they were only manipulating genes!) are now redefining themselves as sheep, and green sheep at that.'

6

Modern Science of Agriculture

THE ENVIRONMENTAL CRITICS

Although environmental philosophers have had little to say about agriculture, environmental critics have not been so reticent. Indeed, the volume of criticism has been so great that it is impossible to even summarize it in less than encyclopedic terms. Critics have found problems with virtually every element of agricultural production and food processing, from centre pivot irrigation to the use of antibiotics in animal feed.

Since a thorough review of these criticisms is out of place in this context, it will be necessary to select a few examples that illustrate how critics have interpreted the environmental implications of agriculture. Criticisms of agricultural pesticides and of emerging agricultural biotechnologies tend to cite a laundry list of negative environmental impacts associated with agriculture. To the extent that this pattern of criticism is typical, it has three important implications.

First, the pattern of criticism makes no philosophical distinction between risk to humans and risk to non-human animals and ecosystem integrity. It is, for this reason, somewhat retrograde by the standards of environmental ethics. Second, by stressing unwanted outcomes, the critics unintentionally reinforce the dominant philosophical orientation of modern industrial agriculture.

Finally, the pattern invites farmers and agribusiness to respond by ameliorating practices, rather than by undertaking fundamental reforms. The review of critical literature begins by noting how environmental critics of agriculture are situated within a four way network of critics who have besieged agriculture since World War II. The interests represented by the three other groups of critics overlap, but do not coincide with those of environmentalists.

Environmental criticisms have been, therefore, diffused by the complexity of messages registering in the minds of agricultural leaders, and have never been interpreted as calling for major changes in the value systems that undergird agriculture in the United States, Canada, Western Europe, Australia, India and other centers of agribusiness and industrialized production. The mixture of messages outlined here pervades the discussion of alternative philosophical strategies for addressing environmental problems in agriculture, and complicates the task of formulating an environmental ethic for agricultural production.

Selective reviews of chemical insecticides and of new technologies based on the transfer of genetic materials through recombinant DNA follow the overview of agriculture's critics. The purpose is to provide paradigmatic examples of environmental criticism. Briefly, the vast literature of criticism provides surprisingly little that would lead to a philosophically novel approach to the environmental significance of agriculture. The final section briefly reviews the comments of two well-known critics, Aldo Leopold and E.F. Schumacher, who do provide preliminary sketches of what an environmental ethic should achieve.

THE CRITICS OF AGRICULTURE

The recent history of agriculture in industrialized countries is a history of technological change. Machines, chemical inputs, and genetic improvements were developed at an advancing pace throughout the twentieth century, and many of these technologies were widely adopted by farmers in industrial countries.

During the 1940s, a group of scientific, business and political leaders conceptualized the Green Revolution, a massive effort of technology development that was to reproduce what they took to be the success of industrialized agriculture in developing countries.

Percentage of chronically Malnourished Population has Declined in Most of the Developing world, Except sub-saharan Africa

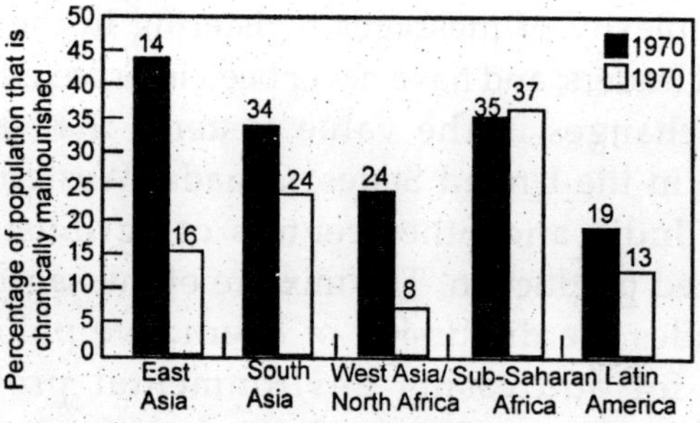

SOURCE; Gondon Conway, The Doubly Green Resolution Food for All in the 21st Century, Ithaca University Press, 1998

The Green Revolution strategy of aggressive applied scientific research, followed by equally aggressive efforts of technology transfer, was intended to improve agriculture by improving the tools and materials of farming. It was a distilled version of the philosophy behind the laws that established the agricultural experiment stations in the United States in 1887, and the state and national agricultural extension services in 1914.

The two laws placed land grant universities in each state in harness with research and extension efforts at the US Department of Agriculture and created what has come to be known as the USDA/land grant system. The rationale for these laws was the general betterment of rural communities, but they have gradually been implemented in ways that focus ever more narrowly on technology.

Whether one speaks of the USDA/land grant philosophy that gave rise to agricultural technologies in the United States or the Green Revolution philosophy that spread a somewhat narrower view of technology transfer across the globe, agricultural science is linked to technology adoption by farmers. In many cases, the technology eventually adopted by farmers must be supplied by firms that manufacture the tractors, cultivators, chemicals, or seed varieties that may have been originally researched in universities or public agencies such as the Agricultural Research Service.

Therefore, the recent history of agriculture is also the history of emerging agribusiness firms that supply farmers with the technologies needed for food and fiber production. The transformation that has taken place in agriculture is dramatic. When industrialized production systems are measured in terms of productivity, the success of this philosophy is startling.

Farming technology has increased the productivity of agriculture, and the result is that people in industrial countries expend far less of their income on food and fiber than do those in countries where farming continues to be the primary occupation of most people. Most participants in the USDA/ land grant, Green Revolution complex (including scientists, public officials, agribusiness firms, and adopting producers) would regard these changes as a successful application of technology, but a chorus of voices has arisen in criticism.

The primary purpose of this chapter is to examine the environmental critics, but it is helpful to realise that environmental criticisms of agriculture are but one voice in a four part harmony of critique. Perhaps the first themes were sounded by critics of the Green Revolution itself.

According to DeWalt, Carl Sauer wrote as early as 1941 that Mexican agriculture cannot be pointed toward "standardization of a few commercial types without upsetting native economy and culture hopelessly." Published critiques of the Green Revolution began to appear in the late 1960s and early 1970s, with Keith Griffin's book, *The Political Economy of Agrarian Change: An Essay on the Green Revolution,* being the first in a series of studies that documented the social turmoil associated with agricultural technology transfer in developing countries.

It distorts the criticism of the Green Revolution to generalize, but it is fair to say that critics have consistently applied two themes. First, critics reject the assumption that changes in agricultural production technology can be evaluated in terms of aggregated indicators such as increased food production or total rural income. They insist that inequitable distribution of benefits and harms overturns the judgment that agricultural technologies have produced success in the developing world.

Second, the presumption that scientific research can produce beneficial changes in any cultural and political environment is replaced with the view that Western science is deeply dependent on the social institutions of developed economies. By altering these two presumptions of the Green Revolution, critics conclude that the loss of local autonomy and indigenous knowledge far outweighs any benefit from increased agricultural production. Frequently, critics apply revolutionary political rhetoric in placing the ultimate blame for Green Revolution failures on capitalist or free market ideology.

A second group of critics have noted the social implications of agricultural technology within developed economies. In short, critics think industrial technology is inimical to small farms and rural quality of life. The origin of this theme may be a study by anthropologist Walter Goldschmidt. Originally published in 1947, Goldschmidt's *As You Sow* prefigured many criticisms that were to mount in the following four decades, as family farmers in rural communities became increasingly aware of their plight.

The attack was leveled directly at USDA/land grant organizations in an influential 1973 book, *Hard Tomatoes, Hard Times* by Jim Hightower, who was later to become Texas Commissioner of Agriculture. Hightower's book and a series of papers on the development of the mechanical tomato harvester in California laid the responsibility for the lost employment and farms that ensued at the doorstep of the University of California. The result was a lawsuit, filed on behalf of displaced field workers, that was settled in favour of the University in 1991.

The general theme of domestic critics is a populist one. Government, including the USDA/land grant system, should protect the "little guy" from the forces of impersonal industrialization, in general, and from big business in particular. The trend in agriculture has manifestly been toward larger, more specialized farms and farm input businesses, and this trend is antithetical to the populist ethic of family farming.

These critics often mix environmental critiques with their populist themes, so Marty Strange's book, *Family Farming: A New Economic Vision* incorporates the critique of centre pivot irrigation alluded to earlier. The environmental dimension of populist critique will surface in later, but one must assess these critics as being *primarily* concerned about the loss of small farms, and generally of the opinion that, if small farms were preserved, the environmental problems of agriculture would take care of themselves.

The final voice of criticism is the most obscure, and is, like the lowest of bass notes, indiscernible to the casual listener. It is, however, arguably the most effective in influencing the direction of change within agricultural institutions. In 1972, the National Research Council (NRC) of the National Academy of Science issued what has come to be known as the "Pound Report." The Pound Report took agricultural universities and the USDA to task for not being scientific enough. The substance of the report attacked needless replications and duplication of studies from state to state, and noted that many agricultural scientists appeared to be working on subjects of little scientific merit or interest.

Other NRC reports have been critical of agricultural universities, including the 1989 report, *Alternative Agriculture,* which criticized USDA/land grant administrators for failing to investigate alternatives to the mechanically and chemically intensive technologies of conventional agriculture. While there has been grousing about the accuracy of NRC studies, they have arguably done more to promote change in agricultural institutions than have all the other critics combined.

The philosophical importance of NRC criticisms is quite different from those of the Green Revolution, which stress equity and autonomy, or of the populists, which stress family farms. The NRC reports question the academic or scientific integrity of the agricultural research system, arguing that the methods for identifying research priorities and funding agricultural science continue to be too much influenced by parochial and non-scientific interests.

Primary among these interests are agribusiness firms. Chemical companies, for example, fund many graduate assistantships to perform blind tests on the efficacy of new pesticides. Such tests are of no scientific interest. They may be limited to blind data collection which renders them useless even as learning experiences. However, the family farmers, Green Revolution critics, and environmentalists who also seek to influence agriculture and agricultural research are no less parochial and non-scientific than agribusiness.

The National Research Council would no more like to see the research agenda in agriculture controlled by the Sierra Club than by the Monsanto Company. As such, though NRC criticisms arc effective in changing research directions, their effectiveness is only accidentally related to environmental issues.

The final point before considering some environmental critics in more detail is to note that the choir of critics makes the evaluation of agriculture and its impacts exceedingly complex. Environmental criticisms can run badly afoul of equity concerns in agriculture. Each of these themes is sometimes reinforced, sometimes undercut by critics who want agriculture to be more scientific. Agricultural ethics is always an exercise in balancing these multiple themes against one another, and there are seldom only two sides to an issue.

The Critics of Pesticide

The implicit ethical basis for agricultural research in the USDA/land grant system is *utilitarian* in that it defines the value of research in terms of its capacity to improve the balance of costs and benefits associated with agricultural production. It is also *anthropocentric* insofar as the balancing act is limited to costs and benefits to human beings.

According to this implicit ethics, the most fully justified research project is the one that promises to achieve the greatest good for the greatest number of people. When distributional issues as well as costs to non-human animals and the environment are ignored, the utilitarian view of social ethics makes it easy to think of ethical evaluation as a form of calculation in which all the benefits and costs of various options are weighed, because the ethically justified course of action is that which best satisfies the rule of maximizing aggregate good (or in situations where all options are unattractive, of minimizing evil).

Even when distributional issues and costs to non-human animals and the environment are bracketed, however, there is one general problem with optimizing decision procedures such as the utilitarian ethic described above: it is impossible in practice to obtain complete and reliable information on all the relevant consequences of a policy decision. Some factors are inevitably left out, and when these factors affect human health and safety, the economic well-being of minority groups, or the quality of the human environment, the entire moral calculation of relative benefits can be drastically skewed. Costs or harms that are simply left out of a utilitarian calculation can be called *externalities*.

Costs are sometimes left out because the decision maker does not have a reliable way to measure or compare them, and costs are also left out when persons or groups deciding on behalf of their own interests do not have to bear them. From the point of view of the self-interested decision maker, the costs are truly external. Decision makers entrusted with the public good must make stringent efforts to reflect all such costs in any estimate of total social benefit or harm, or they cannot truly be said to have optimized outcomes.

The majority of environmentally based criticisms of agriculture and agricultural research clearly take the form of noting the factors and impacts that have simply been left out of the assessment of costs and benefits. The most celebrated of all works in the critical literature, Rachel Carson's *Silent Spring,* follows the pattern of citing externalities by providing an extensive list of unintended consequences associated with the use of insecticides.

Carson was a gifted writer with a flair for evoking the beauty and dignity of wildlife and natural habitat and for expressing outrage at practices that place nature at risk. Nevertheless, her basic philosophical strategy in the book was simply to identify unintended and negative consequences. *Silent Spring* is a laundry list of unwanted consequences and risks that had not been accounted for in making assessments of pesticide use in agricultural production.

Silent Spring is just the first in a long line of critiques identifying environmental impacts of chemicals used in agricultural production. Carson's arguments were reiterated and extended in Frank Graham's 1970 book *Since Silent Spring* and in Robert van den Bosch's *The Pesticide Conspiracy,* to cite two of the most prominent titles. While Carson noted toxic effects upon non-target species, Van den Bosch identified unintended consequences that typify the class of ecosystem outcomes of intense interest to environmentalists.

According to Van den Bosch, chemical pest control ignores ecological forces that control insects. The number of insect species classified as pests doubled from 1962 to 1978, despite increasingly efficient chemical control, skyrocketing insect control costs, and worsening environmental impacts, a phenomenon that Van den Bosch labels an "insecticide treadmill." Insecticides or biocides kill natural enemies of insect pests, and eradicate the natural predators and parasites. A biotic vacuum is then created where the surviving pests thrive without predators or parasites.

Continued spraying becomes a necessity. Van den Bosch makes an explicit ethical argument in his book when he indicts chemical company salespersons and advertisers for their effect upon a farmer s decision to use more pesticides and contribute to the treadmill. Scientific societies and administrators of land grant universities, where the nations pest control research is conducted, are also implicated when pressures and political reprisals from chemical companies affect USDA/land grant research programmes. Van den Bosch's condemnation of public sector agricultural research is based upon a judgment that administrators and scientists have been "captured" by the commercial pesticide industry, and have thus failed in an ethical obligation to conduct research in the public interest.

His criticism is still consistent with Carson's original way of identifying uncounted costs, however, for Van den Bosch clearly understands the public interest in terms of maximizing benefits for human beings. The problem he has with chemical insecticides is that their costs outweigh their benefits. It is profitable for chemical companies to sell pesticides and for farmers to use them only so long as important long-term costs are not included in the overall assessment (or, to say the same thing, so long as the costs are "externalized").

In this case costs are externalized either in the sense that they are borne by individuals whose interests are not included in the tabulation of consequences or in the sense that costs occur beyond the time frame for which consequences have been assessed.

Van den Bosch's criticism of insecticide came forward within a complicated political context. On the one hand, concern about the human health effects of pesticide had become commonplace among urban consumers. Publications from the Rodale Press sounded the themes that pesticide-free foods were better for consumers, and that organic production was a realistic possibility. Opposition to pesticide use modeled on *Silent Spring* was expressed routinely in popular articles on environmental themes.

Pesticide had become emblematic of what was wrong with the culture of consumption. Van den Bosch's criticisms, on the other hand, spread a different anti-pesticide message that began to be internalized by entomologists. Working with Texas cotton producers, Perry Lee Adkisson and Ray Frisbee documented the phenomenon of acquired pesticide resistance in the field. Texas growers using heavy sprays to control the boll weevil unintentionally created a new pest when the pink boll worm, previously not a serious problem in Texas, became resistant to the chemicals being used on cotton. They were among a group of agricultural scientists who began to develop a strategy called Integrated Pest Management (IPM). Opposition to pesticides oscillates between the two poles of critique represented by Carson and IPM.

Many recent critics of pesticide have further modified the claims of Carson, Graham, and Van den Bosch, but they continue to list a broad range of unwanted consequences from chemical agriculture. David Pimentel has produced a series of papers written with a variety of coauthors documenting unwanted effects of pesticides. Pimentel s early work was based on an energy audit of pesticide use, noting that the energy used in the manufacture, transport, and application of chemicals severely compromised the energy efficiency of farming.

He has documented the growth in consumption of pesticides, collected citations of scientific studies indicating the risks to human health, and mounted an argument for drastic reductions in pesticide use. Another stream of criticism notes the use of pesticides in developing countries. Pesticides long banned in the United States were used extensively in developing countries for many years, and continue to be used in some applications.

Critics have argued that the export of these banned chemicals causes significant human health risk to agricultural workers in other countries, and eventually to developed country consumers, who consume fruits, coffee, and other products that may contain residue of long banned chemicals. When continued use of long banned chemicals is factored into the argument, the list of unwanted impacts from pesticide use cannot fail to impress one with the continuing seriousness of pesticides' unintended consequences.

Advocates of IPM stress a much narrower range of unintended consequences, so much so that they represent an alternative to the Carson-Pimentel line of argument. The IPM story deserves attention from anyone contemplating an ethic of the environment. Insect life in farm fields is a model in miniature of wildlife ecology. Some insects feed on plant matter; some are predators that feed on other insects. Insects become agricultural pests only when they do economic damage to crops, and this happens only when their numbers are not sufficiently controlled by predation.

Insects that feed on pests are called *beneficial insects,* so the farm field is an ecosystem where pest and beneficial insects stand in some form of balance with respect to one another. As pest populations grow, food for beneficial insects is easier to find. Eventually, the population of beneficial insects will grow in response to the easy pickings, and the number of pests will decline.

The balance between pest and beneficial insects is far from perfect from the farmer's perspective, but there are, in any given field, likely to be a few species that feed on the crop, but which do not become serious pests in virtue of the fact that their populations are controlled effectively by beneficial insects. Chemical pesticides are not selective; they kill pest and beneficial insects. After a pesticide application has lost its effectiveness (usually within a few days or weeks), insect populations begin to rebound. For plant eaters, there is plenty of food, so population builds quickly. Beneficial insects cannot begin to rebound until there is an adequate amount of prey. This creates a window of opportunity for insects normally controlled by beneficials. With their natural enemies in disarray, their populations can grow rapidly.

Although the balance between pests and beneficials will eventually be restored, farmers can expect a surge of pest insects that will take place before beneficial populations have an opportunity to rebound. Given the pesticide practices in use prior to 1980, farmers would notice the surge and make a new application of pesticide. Beneficial insects never had an opportunity to rebound.

Pesticides are costly to buy and to apply. Entomologists began to discover that sometimes farmers could do better financially by accepting crop damage from pests than by getting on the pesticide treadmill. The treadmill phenomenon becomes even more serious when an insect acquires resistance to chemicals being applied. Pesticide resistance is a textbook example of natural selection. When pesticides are applied, a few insects in the population may be resistant to their toxic effects. These insects will constitute a much higher percentage of the total population after spraying than before. The reason is obvious enough: most of the non-resistant insects are dead.

If the farmer sprays again, the percentage of resistant insects will be greater still. As their percentage grows, these insects begin to pass resistance on to subsequent generations. Under the selection pressure of toxic chemical sprays, insect populations can acquire widespread resistance to the toxic effects with surprising rapidity, rendering the pesticide useless.

Now, IPM entomologists do not reject the use of pesticides. The IPM philosophy holds that on some occasions chemical use is economically viable, and even necessary Pesticides will only be effective at these needed times when insect populations have not become resistant to them. Minimizing pesticide use limits the number of resistant insects in the population, and can significantly extend the amount of time that it takes for a population to become resistant to a given chemical compound.

This is an extremely significant fact for ethics, for it converts IPM from being a simple norm of financial prudence to being a general social norm for farmers of a given crop. The reasoning for this conclusion deserves careful attention even by those who reject the premise that pesticides will continue to be useful and necessary. In the first exposition of the pesticide treadmill, a farmer is wise to limit spraying when the cost of chemicals exceeds the value of the crop protected, particularly when successive sprays will be needed. On this level, IPM is just good business sense. A farmer who wants to waste money with multiple sprays is foolish, but there has never been a moral injunction against foolishness, at least as long as it is one s own money that is being wasted. The decision to use IPM has social implications because even the farmer who practices IPM will want the pesticide to remain effective against pests so that it can be used when necessary.

If the IPM farmer has neighbors who spray wastefully, the insect pests in the region are likely to develop resistance, anyway. Insects that become resistant in the fields of the wasteful will find their way to the practitioners of IPM. Avoiding pesticide resistance requires participation in IPM by all (or most) farmers. It therefore becomes plausible to say that farmers should participate in IPM out of a moral duty that they have to their neighbors.

While it has become difficult to find an entomologist who will not privately confide that farmers collectively use too many pesticides, many continue to reject the stronger claims of Rachel Carson and her heirs. It is the IPM view that dominates among agricultural scientists. Supported by a series of scientific studies, agricultural scientists generally question the seriousness of human and environmental health risks associated with chemical pesticides.

The recent work of Bruce Ames is frequently noted. Ames has discovered that food crops naturally contain a complex mixture of mutagens as part of their natural defence mechanism. He has argued that any cancer risk associated with pesticide residues is overshadowed by risks from these naturally occurring substances. This is not the place to undertake a discussion of whether pesticides do or do not cause harm to humans, wildlife, or other ecosystem impacts. Whether they do or don't is an empirical question, not a philosophical one. However, the pesticide controversy does have philosophical implications.

In framing the pesticide issue as they have, Rachel Carson's heirs make an argument that depends entirely on the factual accuracy of the allegation that pesticide does or may cause harm. This claim admits of three possible responses. One is to deny the accuracy of the claim. A second is to accept the accuracy of the claim, and to look for new technology that mitigates the risk. The third is to accept the accuracy of the allegation, but to argue that both risks and actual harms are outweighed by the benefits of pesticide use.

The first two responses are scientific and technological; they do not raise any ethical issues at all. The third response points toward difficult questions of acceptable risk, but even these questions are entirely consistent with the utilitarian framework discussed above. Risk issues, furthermore, are often dominated by empirical efforts to ascertain accurate measurements of the probability of harm.

None of the responses to critics involve the defenders of pesticide in serious reflection on the values and goals appropriate to agriculture. The pesticide controversy is preoccupied with empirical questions, and has failed to generate much discussion that is fruitful for environmental ethics. If critics of pesticide have hoped to draw upon farmers' sense of moral responsibility for the environmental implications of agriculture, they have failed spectacularly.

CRITICS OF AGRICULTURAL BIOTECHNOLOGY

Chemical pesticides represent an ideal case study for environmental criticisms of agriculture. The case against pesticide is largely an environmental one. Pesticides have not been prominently implicated as technologies that contribute to social dislocation, so populists have given pesticide only incidental attention. Although entomologists doing pesticide work are vulnerable to some of the NRC criticisms, it is also true that several entomologists, including Perry Adkisson, have attained a high level of recognition among scientists for their work on acquired resistance and IPM. Critics of the Green Revolution philosophy certainly object to the use of pesticide in settings formerly committed to peasant subsistence farming, but irrigation, fertilizer, new seeds, and mechanization have been more prominent than pesticide in technology transfer.

The case against pesticides has been made largely by environmental critics, and one would expect that it should provide the clearest signal for identifying environmental criticisms of agriculture.

Biotechnology, by contrast, presents a tangled jumble of criticisms. For one thing, although recombinant techniques for moving genetic material are becoming commonplace in agricultural research settings, few technologies developed through the use of recombinant DNA are currently used in farming or the food industry. One success has been the commercial use of a genetically modified organism (GMO) that produces a very pure form of rennet, a substance used in cheese making. Since rennet has historically been harvested from the entrails of slaughtered calves, the modified bacterium that produces pure chymosin, the active enzyme in rennet, has produced few opponents.

Other agricultural biotechnologies have produced a firestorm of criticism, however. Jack Doyle produced an extended environmentalist critique of biotechnology in his 1985 book, *Altered Harvest.* Doyle described how plant breeders had produced varieties of maize that shared a genetic trait called Texas T cytoplasm. The varieties were planted extensively across the United States, and in 1972 a virus emerged that attacked plants sharing this trait. The result was a disaster for the US corn crop. Doyle used the case to illustrate why it is important that agricultural crops maintain a diversity of genetic traits, and suggested that one risk of agricultural biotechnology would be to increase the chance of a repeat performance.

Defenders of crop biotechnology take the thrust of the criticism seriously, but argue that recombinant techniques give them greater ability to minimize the probability that an entire crop would be susceptible to a given disease. Texas T cytoplasm was produced through conventional crosses that transferred a package of traits, some beneficial, some not, into many varieties of maize. Plant scientists argue that recombinant techniques would have allowed them to home in on beneficial traits with more precision, thus producing more genetic diversity rather than less.

This pattern of claim and counter claim, already evident in the debate over pesticides, also characterizes debates over GMOs. By far the most debate over biotechnology has revolved around recombinant bovine somatotropin (BST), a growth hormone used to increase dairy production. The debate over BST is complex, and serious students of agricultural biotechnology will want to examine it in some detail. BST, however, has not aroused the ire of environmental critics. The main point of contention has been the effects of the new technology on small dairy farms, a theme associated with populist, rather than environmental critique.

Environmental themes have figured in discussion of two technologies, ice-nucleating bacteria and herbicide tolerance, however, and it is worth reviewing each in more detail. By 1981 the commercial potential of gene transfer for agriculture had begun to be recognized. By 1983, one of the first products of these new techniques, the "ice minus" bacteria, was ready for field testing. Intended to inhibit the growth of ice-nucleating bacteria by crowding their ecological niche with genetically altered competitors, the "ice minus" bacteria were expected to extend the growing season for a variety of crops, including potatoes and strawberries, by reducing the likelihood of a crop loss due to freezing temperatures.

At the time that field tests for "ice minus" were proposed, regulatory authority for release of genetically engineered organisms was thought to reside in the National Institutes of Health's Recombinant DNA Advisory Committee (RAC).

The RAC was the outgrowth of a decade of concern over potential health effects of genetically engineered organisms, but the "ice minus" case was novel in important respects. NIH involvement in regulation of recombinant DNA experiments was a legacy of the moratorium on gene transfer of the early 1970s, the Asilomar conference of 1974, and the stringent guidelines for labouratory research that had been established in its wake. By 1983 experience with recombinant DNA had allayed many fears, and NIH guidelines had been successively weakened. NIH had become comfortable with the vast majority of ongoing basic and biomedical research.

Indeed, many of the RAC s most difficult cases dealt not with *safety* of gene transfer but with ethical questions such as the permissibility of altering the human genome. The "ice minus" experiment, however, deviated from the basic and biomedical types of research over which the RAC was understood to have clear regulatory authority. It was also, rightly or wrongly, among the first recombinant DNA experiments thought to have potential for unwanted environmental consequences. By late 1983, then, the regulatory authority of NIH had been questioned, both in the courtroom and by the Environmental Protection Agency.

The saga of "ice minus" grew increasingly complex. The original experiment, proposed under the auspices of the University of California, was blocked by Judge John Sirica on May 16 1984, who in the same ruling held that a private company proposing the same experiment would not be bound by laws requiring environmental impact analysis. Within two weeks of Sirica's ruling, the RAC recommended approval for UC scientist David Lindow's experiment, this time submitted under the auspices of Advanced Genetic Resources, escaping the force of Sirica's ruling. This action merely precipitated a lawsuit, delaying the experiment again.

Both suits were eventually resolved in a manner that permitted the "ice minus" experiment, but NIH also came under sharp criticism in one decision for failing to "sufficiently analyze the potential for the bacteria to be used in the California experiment to disperse or survive in the environment". The "ice minus" experiment had by now received enough publicity to generate public opposition at the strawberry test site near Tulelake, California. County boards in both communities voted to prohibit the experiment, and the California Superior Court issued a restraining order on August 6 1986, delaying the experiment until spring 1987. Delayed by four years, the "ice minus" field test commenced on April 23 1987 at a third site near Brentwood in Contra Costa County.

Regulatory policy for agricultural biotechnology made little progress in the meantime. The overlap between NIH and EPA continued throughout 1984, as environmental scientists stressed the need to assess ecological risks before permitting release of engineered organisms. By 1985, EPA had conducted its own favourable review of the "ice minus" experiment, but regulatory confusion had only deepened as the US Department of Agriculture and the Food and Drug Administration were proposed as addition-al partners with NIH and EPA in a "Biotechnology Science Coordinating Committee."

By 1986, the biotechnology industry itself had begun to call for government involvement in the regulatory process, partially as a way to stifle opposition arising from unwarranted fears and speculations, and the Reagan White House announced a plan that supported NIH guidelines but transferred regulatory authority to EPA and USDA. The new guidelines established the principle that risks of genetically manipulated products should be evaluated on the basis of product characteristics, not manufacturing processes. No sooner had these guidelines been proposed than they were the target of yet another lawsuit.

In October 1986, *Science* reported significant inconsistencies between EPA and USDA, with ecologists expressing concern about the government s intention categorically to exempt certain types of genetically created organisms from environmental impact assessment. In the following month there were similar conflicts between USDA and NIH. Since the furor over "ice minus," government oversight of agricultural biotechnology has increased, but confusion over requirements and regulatory authority still occurs.

At present, EPA and USDA continue to claim authority to regulate recombinant organisms that may affect the environment. Working relationships among agencies have been facilitated through an Agricultural Biotechnology Regulatory Advisory Committee (ABRAC). The ABRAC itself makes no regulatory decisions, but advises several agencies on policy, and by doing so functions as a focal point for coordination. Even within USDA, however, regulatory authority is not always clear.

USDA has an Office of Biotechnology, whose primary purpose has been to foster development of recombinant techniques for agriculture, and a National Biotechnology Impact Assessment Programme, designed to develop procedures for risk analysis. Neither has regulatory authority, however. The Food and Drug Administration (FDA), the Animal Plant Health Inspection Service (APHIS), and the Food Safety Inspection Service (FSIS) all have authority, but would not normally be concerned with research, such as was involved in the "ice minus" case.

Research sponsored by USDA through the Cooperative State Research Service (CSRS) is regulated as part of the proposal review process, and universities that conduct publicly funded research are required to have an Institutional Biosafety Committee (IBC). The IBC will normally review all research involving recombinant DNA, yet it remains the case that privately conducted research is not subject to regulatory review. To sum up, regulatory policy for agricultural biotechnology lies buried deep in the forest of government acronyms.

A venture into the woods may or may not turn up a clear answer as to what is permitted, or when risks are too great. When regulatory authority is unclear, cases ultimately wind up in the courts. For activists such as Jeremy Rifkin, the goal of a lawsuit may be simply to slow down experiments so that affected parties have adequate opportunity to ensure that their interests are adequately protected. The success of litigation initiated by Rifkin's Foundation on Economic Trends (FET) in opposing the "ice minus" experiment complicated that case with jurisdictional issues, but the underlying ethical issue was acceptable risk.

With respect to "ice minus" bacteria, the question had little to do with whether the experiment poses a serious hazard; every review of the proposed research had concluded that it did not. The FET lawsuits, however, exposed a general confusion over what, in fact, is meant by "acceptable risk" and whether one agency's judgment of acceptability is binding on another. The series of lawsuits represents a classic use of uncertainty arguments to raise a succession of doubts, first about the safety of a practice itself, then about the reliability of methods for assessing risk, and finally about the integrity and reliability of experts conducting the analysis.

At the time of the "ice minus" case, regulation of agricultural biotechnology had not found a way to manage uncertainty arguments. In the minds of citizens and affected parties, uncertainty escalates the risk of new products and procedures. The problem is typically described as one of managing public perception of risk, but this is misleading. It is the public perception of the scientific community (and of their methods for analyzing technology's unwanted consequences) that was the basis for the judgment that risks were unacceptable in the "ice minus" experiment.

While the scientific community focuses intently on the characteristics of the organism itself, the public, with little basis for making a judgment on the probable consequences of deliberate release, focuses instead on the characteristics of the scientists. The potential for unwanted impact due to field testing and commercial use of modified plants has been taken relatively seriously by agricultural scientists, though not seriously enough in the eyes of critics.

As the debate has moved on to the technical provisions of regulations and protocols for field testing, and commercial release, the issues have become too arcane for general public consumption. Like pesticide questions, these issues depend heavily on the measurement and weighing of risks, and in balancing risks against expected benefits. The difference of opinion between advocates of biotechnology and opponents boils down to differing estimates of the probability and degree of harm, and value of compensating benefits. This difference of opinion depends largely on issues that can only be settled by empirical inquiry. It is the expense and difficulty of gathering the data to settle the issues that keeps them alive.

The case of herbicide tolerance is quite different. Like insecticides, herbicides typically kill both beneficial plants (e.g. crops) and pests (e.g. weeds). Obviously, this limits the applicability of herbicides rather dramatically, generally to early season use, before crops come up, or to highly targeted use, away from crop roots and foliage. The idea behind genetic engineering of herbicide resistance is that if crops acquire resistance, one can use herbicides with impunity.

Plant scientists have had some success in identifying the genes that code for herbicide tolerance, so recombinant techniques can be used to move these genes to the beneficial crop plants. This is anathema to environmental critics of agriculture, for it seems that biotechnology is being used in a way that will exacerbate the problems of chemical agriculture.

Defenders of herbicide tolerant crops argue that they will allow farmers to apply principles of IPM to weeds, as well to as insect pests. Now, they note, farmers must use herbicides early, before they know whether they even have a weed problem. With herbicide tolerant crops, they could wait until weed infestation threatens to cause economic damage exceeding the cost of spraying before using herbicides at all.

Farmers could quit using the wide variety of herbicides now used that target specific weeds but avoid damage to crops, and could switch to a broad spectrum herbicide that kills all the weeds, but not a genetically modified resistant crop. In addition, of course, they argue that farmers will get better weed control and increase yields. The argument, then, is that contrary to the claims of environmentalists, herbicide tolerant crops may reduce the amount of herbicide used, and will, in any case, produce benefits that more than compensate for chemical risks.

Philosophical dimensions of the case for and against genetically engineered herbicide tolerance have been reviewed thoughtfully by Gary Comstock and a paper by Comstock and molecular biologist Jack Dekker presents the reasoning that led Dekker to discontinue his research in the area. In the present context, a 1990 document, *Biotechnology's Bitter Harvest,* prepared for the Biotechnology Working Group by Rebecca Goldberg, Jane Rissler, Hope Shand, and Chuck Hassebrook is particularly instructive for the way it combines all four voices of criticism noted above.

The report echoes Green Revolution critics by noting Third World impacts of agricultural biotechnology, and cites the NRC report *Alternative Agriculture* in support of its conclusions. The specific complaints against herbicide tolerant crops are noted in a chapter entitled, "The Human Health, Environmental, Social, and Economic Impacts of Herbicides and Herbicide-Tolerant Crops." The chapter text follows the laundry list model implied in its title by listing and documenting a series of unwanted consequences that could follow the introduction of herbicide tolerance into crops and trees.

Among outcomes noted are the suspected carcinogenic properties of specific herbicides, food safety concerns associated with herbicide residues and with the consumption of the modified crops, contaminated drinking water, and interbreeding with other weeds. The emphasis in this list is clearly on health and environmental impact, though questions about a farmer's increasing dependence on private sector technology link herbicide tolerant crops to family farm issues.

The report provides equally extensive listing and documentation of how herbicide tolerance research is organized and funded. Research is underway both within the private sector and at agricultural universities and the USDA. University and government research is funded by a mixture of public and private funding. Funding receives extensive discussion in the report because it is crucial to three central points of criticism. First, the amount of money spent on genetically engineered herbicide tolerance vastly exceeds the amounts spent on sustainable alternatives that would be preferred by the reports authors. The report states, "Perhaps the greatest problem with herbicide tolerance, however, is that it diverts us from paths that really could lead to reduced chemical dependency in agriculture".

Second, public funds are being spent in a manner that effectively subsidizes research costs for chemical companies, or that benefits directly corporations by increasing the market for their herbicides. The implicit premise is that money spent to benefit small farmers would be in the public interest, while money that benefits the input industry does not. Third, the authors argue that chemical corporations are supporting research on herbicide tolerance (both directly and by lobbying public officials) because it helps them gain control of the research agenda in agricultural biotechnology.

While these are all important and interesting arguments in their own right, in this context they are remarkable for the way in which they integrate specific health and environmental criticisms with other forms of concern. Populist themes, in particular, emerge in the implied criticism of links between government and big business. None of the three objections notes environmental impacts. They disparage herbicide tolerance research, but not in virtue of unintended consequences.

Yet, the arguments would have little force *without the* environmental consequences noted above. The authors' review of funding priorities leaves readers with the impression that there is something fundamentally skewed about agriculture and agricultural research. The report is explicit in noting that a turn toward sustainable agriculture would help put things right, and hints that the difference between industrial and sustainable agriculture has something to do with a persons "mindset".

However, there is little more than the list of unwanted outcomes to differentiate sustainable from non-sustainable agriculture. As such, it is far from clear that a truly philosophical shift is what the authors mean by "mindset."

The debate between critics and defenders of herbicide tolerance reprises the debate over pesticides. Despite attempts to bring in "mindset," the facts are what is at issue, and there is little of ethical significance to debate. Again, the form of the criticisms invites defenders to reply in ways that fail to generate philosophical reflection or thought. Furthermore, although the controversy is nominally about agricultural biotechnology, it is really a debate about chemicals. Agricultural biotechnology is being criticized for failing to move agriculture away from dependence on chemical inputs, and for, in fact, offering nothing more than symptomatic relief of the problems engendered by industrial agriculture.

However, if critics express their dissatisfaction in terms of risk to human health, and wild plant or animal species, they should expect a response that focuses on minimizing these risks, or that compensates for harm done. If the problem with chemically intensive agriculture is something *other* than the risks associated with chemical use, critics should not expend so much effort predicting harmful consequences from chemical use. If the problem is that modern industrial agriculture is founded on philosophical premises that are fundamentally flawed, it is those premises that should be exposed and criticized. In short, one should not grumble about responses that address only symptoms when one s original complaint has itself remained at the level of symptoms.

Both *Altered Harvest* and *Biotechnology's Bitter Harvest* merge environmental criticism with some of the populist themes that have been prominent in the attack on BST. Critics create the distinct impression that something is fundamentally amiss in industrial agriculture, and blame the unwanted impacts they cite upon agriculture's dominance by commercial interests, seeking profit from the sale of commodities and from input technologies. Agriculture, however, has been controlled by commercial interests at least since the decline of the feudal system in Europe.

The critics want a return to a more humane agriculture, such as what may have existed in the late nineteenth and early twentieth centuries, but that agriculture was thoroughly commercial. Furthermore, collectivization experiments in the

Soviet Union and other socialist countries have produced both human and environmental problems. As such, the commercial orientation of industrialized agriculture cannot be either a necessary nor sufficient condition for the unwanted impacts of modern agricultural technology, including biotechnology.

Environmental Ethics

The environmental critics of insecticides and agricultural biotechnology provide ample documentation of agriculture's importance for environmental policy. The criticisms they mount, however, do not constitute philosophical problems for agriculture, nor do they represent points of philosophical interest for those constructing an environmental ethic. This is not to say that the environmental criticisms of agriculture have no philosophical implications at all. Indeed, three points of significance illustrate why agricultures environmental impact has been of little interest to environmental philosophers. These points also indicate a line of inquiry for a more philosophical review of agriculture.

First, the critics of pesticide and of agricultural biotechnology recite laundry lists of unwanted impacts, but provide little insight into how or why impacts on nature differ from harm and risk to human beings. The food safety risks associated with chemical residue are as prominent in the environmental critique of pesticides as the impact on wildlife or biological diversity. Setting empirical questions aside, this pattern of criticism neglects a distinction of keen interest to environmental ethicists. One does not need an environmental ethic to explain why harming people are wrong.

An ethic of minimizing suffering or respecting human rights is perfectly capable of accounting for the wrongness of human health risks. It is not clear, however, that traditional ethical theories explain why we should be concerned about impact upon wildlife or biological diversity. Environmental ethicists have dedicated themselves to the task in a manner that is reviewed at several junctures elsewhere in this volume. Far from providing any unique or unifying environmental theme for philosophers to consider, the laundry list style of criticism appears to lack sophistication in themes that have already been well covered by philosophers.

Second, by stressing unwanted outcomes, the critics are working within, rather than against, the existing utilitarian philosophical framework of industrial agriculture. Industrial agriculture is committed to an ethic of optimizing the trade-off between costs and benefits. It has no intrinsic commitment to chemicals or to molecular technology. Business and scientific practices made it easy to overlook some of the costs of pesticide technology, but if they are true to their utilitarian principles, researchers and planners must be cognizant of all consequences.

To the extent that the critics help decision makers attain cognizance, they assist utilitarian evaluation of agricultural practices. They make no philosophical objections to the optimization philosophy of utilitarian agriculture, in any case. Third, the critics implicitly invite farmers, businesses, and researchers to solve environmental problems by developing alternatives that avoid or compensate for unwanted impacts. If the problem with pesticide or agricultural biotechnology consists in unintended consequences, why not keep doing what we're doing, but get rid of the consequences no one intends?

As already noted, the criticism provokes a technical response rather than philosophical reflection on the part of producers or scientists. If environmental critics truly want agriculture to rethink its philosophical bent toward production, as they seem to, they will have to mount an attack that goes beyond a list of unwanted outcomes. Such lists not only serve the existing ethic of industrial agriculture by requesting that producers correct their cost accounting, but they present little of interest to environmental philosophers, who might help envision an alternative agriculture.

Leopold and Schumacher

Two of the formative intellectual figures in the environmental movement, Aldo Leopold and E.F.

Schumacher, did include some discussion of agricultural philosophy in their writings. Leopold is best known among philosophers for his essay "The Land Ethic" from *A Sand County Almanac*. The essay begins with a passage in which Leopold describes the rejection of human slavery as one of the key instances of moral progress in history. The key to this event, he thinks, was in ceasing to understand human beings as property, in extending the scope of the moral community to include all human beings.

Leopold's message is that we must now find a way to think of our relation to land, understood again to mean the general biosphere, as something other than mere property. Leopold finds any attempt to reflect conservationist concerns within the kind of optimizing calculations that underlie a traditional approach to agricultural decision making hopelessly lacking. In Leopold's view there is ample basis for care and concern about ecological values, but the problem is that the importance people place upon nature cannot be reflected in monetary terms. He writes:

> When one of these non-economic categories is threatened, and if we happen to love it, we invent subterfuges to give it economic importance. At the beginnings of the century song-birds were supposed to be disappearing. Ornithologists jumped to the rescue with some shaky evidence to the effect that insects would eat us up if birds failed to control them. The evidence had to be economic in order to be valid.

Here Leopold would also seem to be rejecting the notion that unwanted outcomes of agricultural production decisions can be accommodated by a broader framework of benefits and costs, and including some constraints. Indeed, it is property rights, Leopold's target, that serve as the model for constraints. Instead, we must rethink our lives and our values so as to attain a fuller appreciation of the interdependence between human and natural communities.

One of the chief sources for understanding these links is agriculture. Early on in his book Leopold writes, "There are two spiritual dangers in not owning a farm. One is the danger of supposing that breakfast comes from the grocery, and the other that heat comes from a furnace". One who lives on a farm cannot, in Leopold's view, long forget the dependence of human action upon the underlying natural ecology. Written in the 1940s, *A Sand County Almanac* does not reflect more recent critics' concern that agriculture is on the verge of destroying its ecological base, but Leopold does express cynicism about the optimizing strategies of experiment station research:

The State College tells farmers that Chinese elms do not clog screens, and are hence preferable to cottonwoods. It also pontificates on cherry preserves, Bang's disease, hybrid corn, and beautifying the farm home. The only thing it does not know about farms is where they came from. Its job is to make Illinois safe for soybeans.

Leopold's land ethic, thus, rejects the optimizing strategy that takes increasing income, increasing production, and increasing benefits to consumers as its core. Instead, Leopold urges us to:

Examine each question in terms of what is ethically and aesthetically right, as well as what is economically expedient. A thing is right when it tends to preserve the integrity, stability, and beauty of the biotic community. It is wrong when it tends otherwise.

E.F. Schumacher's 1972 book *Small is Beautiful* followed *Silent Spring by* a decade, and combined Carson's concern for agricultural technology with Leopold's s distaste for making moral evaluations by calculating costs and benefits. The central theme of the book was widely taken to be an attack upon technologies that consumed relatively large quantities of fossil fuels and required large investments of fixed capital. However, in the chapter entitled "The Proper Use of Land," Schumacher takes up a central question in agricultural ethics. The argument of the chapter is first a criticism of what Schumacher calls "the philosophy of the townsman" and, second, a description of an alternative programme.

The "townsmen" see agricultures economic woes as evidence that farming or ranching is a declining enterprise, and see the central problem of agriculture as one of improving farm income. Schumacher finds this view deficient. He writes:

We know too much about ecology today to have any excuse for the many abuses that are currently going on in the management of the land, in the management of animals, in food storage, food processing, and in heedless urbanization. If we permit them, this is not due to poverty, as if we could not afford to stop them; it is due to the fact that, as a society, we have no firm basis of belief in any meta-economic values, and when there is no such belief the economic calculus takes over.

In Schumacher's view, the problem arises when agriculture is understood as essentially defined by its capacity to produce and market saleable commodities. In making a statement of the wider goals of agriculture he writes:

A wider view sees agriculture as having to fulfill at least three tasks:

- To keep man in touch with living nature, of which he is and remains a highly vulnerable part;
- To humanize and ennoble man's wider habitat; and
- To bring forth the foodstuffs and other materials which are needed for a becoming life.

I do not believe that a civilization which recognizes only the third of these tasks, and which pursues it with such ruthlessness and violence that the other two tasks are not merely neglected but systematically counteracted, has any chance of long-term survival.

These remarks on agriculture must be understood in the light of Schumacher's overall attack upon "economic values" and his campaign to substitute a norm of "Buddhist economics" in its place. In criticizing economic values Schumacher means to attack the utilitarian emphasis upon increasing incomes; by interposing "Buddhist economics" in place of this emphasis, he means to suggest that there is an alternative way of conceptualizing economic activity, one that would trace production, distribution, and exchange according to the long-term impact of these activities upon the natural systems needed to support all.

Economic policies that encourage consumption in order to promote economic growth are, on the view of Buddhist economics, incompatible with the goal of a permanent and stable society. Although Schumacher's choice of words has the ring of late sixties hippie jargon, his point should be understood as a shift in philosophical perspective. Political theorist Paul Diesing has argued that Schumacher's critique is a complete rejection of the traditional utilitarian perspective on agricultural production. On this traditional view, Diesing writes:

> Nature appears in three forms: natural resources, cultivated land..., and externalities of production. Natural resources are free goods, *res nullius,* nothings, having no value until they are "produced" and made available for exchange.

When the central goal of agriculture is understood in terms of production, agricultural land is a form of fixed capital, and this, in turn, suggests that this land should be devoted to its most productive use. Although unwanted outcomes can be factored into the optimizing equation either as costs or as constraints, the result looks a bit like pre-Copernican models of the solar system, where epicycles and reversing rotations were added on to the charts for planetary motion in order to preserve a theory that falsely placed the earth at the centre of the universe.

In Diesing's view, Schumacher rejects this strategy when he insists that agriculture is not a form of industry and should not be viewed as fixed capital or even as a factor of production at all. Instead, land is the basis for life itself, a precondition for productive economic life, and not merely one among many factors available for productive appropriation. In Diesing's view, the agrarian component of Schumacher's thought is its essential philosophical theme. The more celebrated work on appropriate technology flows from Schumacher's view of agriculture, rather than the reverse.

The citations from Leopold and Schumacher indicate how each had a view of agriculture that was inconsistent both with the utilitarian orientation typical of producers or agricultural researchers, and with the presumptions of academic environmental ethicists. They make comments hinting at a philosophy of agriculture that includes an environmental ethic drawn from the very practices of farming itself. The development of such an ethic, and an evaluation of its significance to broader questions in agriculture, environmental policy, and environmental philosophy is the task that now awaits.

7

Green Revolution

CROP TECHNOLOGY

Genetically modified crop technology has revolutionized agriculture in the United States, Canada, China, and Argentina. It exhibits the potential to have much wider impact, solving many of the current problems in agriculture worldwide. The types of GM crops that may become available in the future could boost crop yields while enhancing the nutritional value of staple foods and eliminating the need for inputs that could be harmful to the environment. While the environmental, health, and economic risks of GM crops should be carefully studied before full-scale adoption, the types of GM crops that are already available have thus far largely proven to be beneficial to agriculture and even to the environment, without evidence of adverse health or environmental impacts.

Yet, in other than the four countries mentioned above, the GM crop movement has had little or no impact. In those parts of the developing world where an agricultural revolution might be most welcome, the Gene Revolution has yet to be embraced. Why is this so?

For one thing, the Gene Revolution began in a different way than the Green Revolution. GM crops were first created within the context of the biotechnology industry to provide enhanced agricultural technologies to the industry's primary customers—farmers in the industrial world. These crops were not meant at the outset to be a life-saving technology for the developing world. Although it is almost certainly possible from a scientific and technological standpoint to create GM crops that would be beneficial to developing-world farmers, neither producers (the biotech industry) nor consumers (developing-world farmers) have sufficient economic incentives for this to happen. In fact, the enormous costs of producing each GM crop variety could prove to be a disincentive for the industry to develop "orphan GM crops" that would benefit developing-world farmers.

Additionally, even if the biotech industry were to develop GM crops that are beneficial to farmers in the developing world, the poorest of those farmers would not be able to afford GM crop seed instead of conventional varieties, much less purchase new GM crop seed for every planting season, as biotech patents would require them to do.

Finally, the current political situation is not as conducive to promoting this new agricultural movement as it was for the Green Revolution. For all the potential that GM technology holds, there are many challenges to be overcome if GM crops are to truly introduce a "Gene Revolution" worldwide.

AGRICULTURAL BIOTECHNOLOGY FOR THE FUTURE

Given the challenges stated above, it is important to keep in mind that agricultural biotechnology may not be the best solution, or even a one-shot solution, for all parts of the developing world, for three reasons.

First, as of yet, there are few if any sustainable technological solutions for controlling pests and pathogens in subtropical subsistence agriculture. Currently, in the poorest agricultural areas, food production is feasible only with very low inputs of semi-landrace material of many different genotypes planted together to be broadly adapted to local environments. If one genotype fails, then the others may still succeed on a year-to-year basis, thereby achieving some level of security in the food supply. GM crops, unless they are created from many different hybrids and are modified to withstand a broad range of environmental fluctuations, could not be expected to consistently improve yield if planted alone in subtropical areas.

Second, there are usually alternative ways to conduct public health or agricultural interventions, and all interventions have attendant costs. GM crops may be among the more costly interventions given their current R&D costs as well as the costs to growers. Malnourished people may not need GM golden rice to prevent blindness, for example, and policymakers should first take a step back to see which choices make the most sense in terms of both long-term sustainability and cost considerations. One possible intervention is enhanced conventional breeding. The newest conventionally bred crops have some immunity to common plant diseases and resistance to pests while retaining high yields.

Conventional breeding, while theoretically having far greater limitations than agricultural biotechnology, is less controversial from a global viewpoint and may be less expensive. Hence, in the short term, enhanced conventional breeding may be crucial to improving agricultural yields in areas that do not want to risk losing their food export markets due to current political tensions or government regulations, and it may be important to farmers with limited monetary resources.

Other methods of promoting sustainable agriculture may also prove to be useful—for example, the adoption of farming techniques for greater economic return, such as agroforestry (to increase income), reclamation of degraded land, and irrigation scheduling. As an alternative to introducing GM seeds now, a possible intervention that could be helpful in the poorest nations is the empowerment of women, who are currently the crop harvesters.

For example, they could be educated to become agricultural scientists, learning to select seeds for desirable qualities, such as improved yield and improved quality. This could be a first step toward agricultural independence, which could then make for a smoother transition to agricultural commercialization.

Third, it would be overly simplistic to imagine that improved crop varieties, whether GM or enhanced conventional crops, are all that are needed to ensure food security. It is important to remember that the root cause of hunger is poverty—the inability to access food or the lack of a means to produce it.

Many factors contribute to poverty, not just poor food production. Farmers in the developing world also need support from certain political and social infrastructures that can safeguard incentives to use the GM crop technology appropriately. If the Gene Revolution is to succeed in the developing world, many of those infrastructures must be in place to ensure the long-term benefits from GM crop planting.

Lessons from the Green Revolution

Notwithstanding its attendant challenges and alternatives, the GM crop movement shows great promise. Like the Green Revolution before it, the GM crop movement has the potential to achieve substantial production increases in regions of need and (unlike the Green Revolution) to reduce the need for agricultural chemicals and scarce resources, such as water. Both the successes and failures of the Green Revolution provide useful lessons for how to make GM crop technology a desirable and sustainable agricultural movement in the developing world.

The Green Revolution demonstrates that to create GM crops that are truly beneficial to the developing world, plant breeders and other scientists must be familiar with the local environment and the planting methods of the region for which they are developing crops. Often times, agricultural conditions in developing regions are so different from those in the industrial world that it is difficult for industrial world scientists to know how to devise appropriate technologies for those regions. During the Green Revolution, plant scientists traveled abroad extensively, developing crop seeds that were best suited to particular regions given their particular weather conditions, plant pests, water availability, and planting seasons.

Importantly, these plant scientists trained others in each region to be able to carry out the Green Revolution practices independently. The same sort of global effort is needed for the Gene Revolution to take hold in the developing world. For this global effort to take place, however, there must be a vested interest on the part of those entities that control the Gene Revolution technologies—those that create the technologies, namely those in the biotech industry and those that regulate the technologies nationally and internationally.

The Green Revolution owed much of its success to public-sector institutions that poured resources into the effort, as well as to regulatory regimes in both the donor and recipient nations that were permissive and even encouraged adoption of the new agricultural technologies. Times have changed, though. R&D for GM crops is supported by the public sector only in unusual circumstances, with the biotechnology industry mostly creating GM crops that are beneficial to industrial-world farmers, its primary customers (and those who can afford to pay for the technology).

To complicate matters, current intellectual property regulations that protect the biotech industry's creations limit the flow of information on how to create GM seeds to the public sector, making it difficult to garner the public support needed to develop crops for the poorest farmers in the world. IP rights also lead indirectly to increased GM seed costs that make GM seeds unaffordable to most developing world farmers without significant subsidization. Collaborations between the public and the private sectors to promote the Gene Revolution in the developing world do exist, but thus far only in isolated instances on a small scale.

Further hindering GM crop adoption worldwide is the lack of uniform regulation of foods derived from modern biotechnology. Unlike the permissive regulatory environment of the Green Revolution, in which agricultural advances were encouraged for both philanthropic and political reasons, decision makers today are largely divided into two camps on whether GM crops should flow freely through the food system.

The European Union's new regulations on traceability and labeling of GM foods would require a crop-segregation system that is almost impossible to achieve in a nation without a highly developed commercial agricultural sector.

Thus, developing nations may find it in their best interest to avoid planting GM crops altogether, despite the agricultural and nutritional benefits that GM crops might provide. In addition, many NGOs and other organizations have expressed concerns about the risks surrounding GM crops, and their opinions are becoming increasingly important to the public debate and decision making process. These groups and the average citizen have seen little public benefit from the types of GM crops produced today, except for perhaps slightly cheaper food.

What can we determine about the prospects for the Gene Revolution by studying the Green Revolution's successes and failures? The Gene Revolution thus far resembles the Green Revolution in the following ways:

1. It employs new science and technology to create crop seeds that can significantly outperform the types of seeds that preceded it;
2. The impact of the new seed technologies can be critically important to developing-world agriculture; and
3. For a variety of reasons, these technologies have not yet reached parts of the world where they could be most beneficial.

On the other hand, the Gene Revolution is unlike the Green Revolution in the following ways:

1. The science and technology required to create GM crop seeds are far more complicated than the science and technology used to create Green Revolution agricultural advancements;
2. GM seeds are created largely through private enterprise rather than through public sector efforts; and

3. The political climate in which agricultural science can introduce new technologies has changed dramatically.

The similarities and differences between the Green and Gene Revolutions lead us to speculate that for the GM crop movement to have the sort of impact that would constitute an agricultural revolution, the following goals still need to be met and their related challenges overcome:

Agricultural Biotechnology must be made Affordable to Developing World Farmers

Unless this condition is met, farmers may not see that it is in their best interest to use GM crops, despite the significant benefits those crops could provide.

During the Green Revolution, the new HYV seeds and accompanying chemicals were more expensive than the landrace seeds that developing-world farmers typically had used. Therefore, loan systems and cost-reduction programmes were established regionally in which farmers' eventual profits from increased production could be used to reimburse lenders. In many settings, these programmes proved to be no longer necessary several years after their successful adoption. Current R&D costs for genetically modified seeds are even higher than the R&D costs for the Green Revolution's HYV seeds.

At the price that U.S. farmers currently pay, GM seeds would be unaffordable to most developing-world farmers. Cost-reduction programmes and loan systems similar to those that were established during the Green Revolution must also be established for the Gene Revolution; however, establishing such systems is more difficult now because of higher costs and because the seeds are produced by the biotech industry rather than by agricultural scientists in the public sector.

There is a Need for Larger Investments in Research in the Public Sector

Numerous studies have shown the importance of public sector R&D to agricultural advancements, including the advancements of the Green Revolution. During the Green

Revolution, partly because the R&D and its products were almost entirely in the public domain, intellectual property issues were not a barrier to scientists, for example, taking seeds from one region of the world, hybridizing them with seeds from another region, and producing new seeds to benefit yet another region. Today, however, the production and distribution of GM crops are largely within the domain of the biotech industry, and IP issues are central to the development of GM seed.

While IP laws protect the rights of GM seed creators in industry, those laws are currently an impediment to disseminating the necessary knowledge and technology to those parts of the world that need them. Therefore, public-sector research is essential if the GM movement is to assume revolutionary proportions. Partnerships between the public and private sectors can result in the more efficient production of GM crops that are useful to the developing world and expand the accessibility of those crops and their associated technologies to developing-world farmers.

To Garner the Level of Public Interest and Support that can Sustain an Agricultural Revolution, Agricultural Development must be Regarded as being Critically Important from a Policy Perspective, in both Donor and Recipient Nations.

Without public policy support, cooperation among the many stakeholders in the Gene Revolution will be stymied. For 30 years after World War II, policymakers viewed agricultural development as being essential to world peace. For that reason, policymakers in both the United States and in Asia and Latin America supported the Green Revolution from the start. The end of the Cold War, however, has not brought about an increase in global stability. Whereas the conflict between East and West has declined, there is a growing divide between rich and poor nations.

Unfortunately, with the end of the Cold War, developed nations are concentrating more closely on their domestic political agendas and less on global concerns, and as such have decreased their funding to poorer nations. However, these reductions in aid are not in the best long term interests of even industrialized nations. An increasingly polarized world of the rich versus the poor will result in growing political unrest. Unless developing nations are helped to provide sufficient food, employment, and shelter for their growing populations, the political stability of the world will be further undermined.

As population numbers continue to increase, agricultural development is more necessary than ever to eliminate malnutrition and prevent famine, particularly in sub-Saharan Africa. GM crops are seen as a means for addressing those problems. However, policymakers worldwide are far from being a combined force on this issue; the driving force behind improved agriculture is less unified than it was during the Green Revolution. The question of who should assume the task of re-establishing the importance of agricultural development among policymakers is an issue for further inquiry.

Policymakers in the Developing world must set Regulatory Standards that take into Consideration the Risks as well as the Benefits of Foods Derived from GM Crops.

This goal is crucial to the cooperation of the many stakeholders that are affected by GM crops and also for the sustainability of the GM crop movement in the foreseeable future. A generation ago, the regulatory environment surrounding the Green Revolution was extremely permissive. Scientists could move freely among nations to help breed and plant HYV crops, and there was no stigma attached to eating foods developed from these crops.

Today, however, the regulatory world is divided between those nations that permit GM crops to move freely through their food system (e.g., the United States, Canada, China, and Argentina) and those (primarily the EU) that have strict regulations regarding GM crops in their food systems. There are many possible reasons for the disparity in regulations—differing consumer attitudes, trade issues, and differences in regulatory philosophy among them.

The discord regarding GM crop regulations is currently playing itself out (as of this writing) in a case before the WTO to determine whether the EU's rules on GM foods constitute an illegal trade barrier. In the meantime, policymakers in certain African nations have decided that they cannot afford to permit GM crop planting, even if it is beneficial to their growers and consumers, because they are wary of losing financial aid from the EU if they are seen as taking a proGM crop stance. Without regulations that explicitly take into account potential benefits to both farmers and consumers, those nations that might stand to benefit most from GM crops may be discouraged from allowing them to be planted.

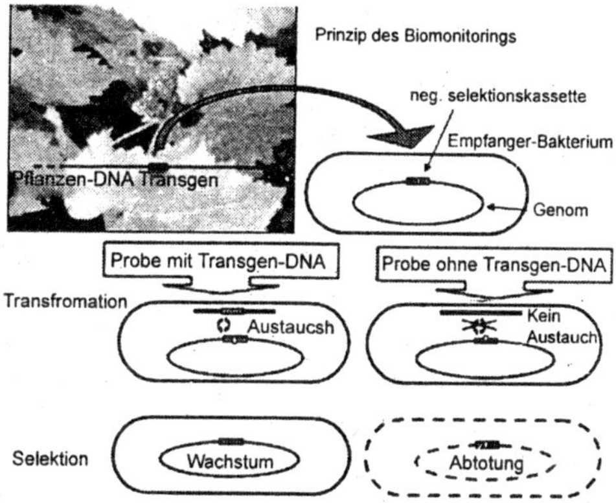

Fig. Transgene outflow

At the same time, policymakers worldwide must ensure that risk assessments of GM crops are conducted to address the specific concerns of their regions. A risk assessment of transgene outflow in the India, for example, is unlikely to be relevant to ecological concerns in Mexico or Africa. In assessing risks, policymakers in developing nations must consider, among other factors, the types of native and agricultural plants that may be affected by the presence of GM crops, traditional farming practices, and the desired traits of GM crops that may be planted in their regions in the near term and long term.

Implications for Relevant Stakeholders

What do these challenges mean for the various stakeholders that are or should be involved in solving the problems surrounding current and future agricultural needs worldwide? First, national governments worldwide should realise that so long as there is any threat of widespread hunger or malnutrition, the threat of political instability and insecurity (partly caused by lack of food security) is larger.

Indeed, problems of hunger and malnutrition still exist, most especially in sub-Saharan Africa, and the benefits of the Green Revolution in other parts of the developing world are slowing. Thus, governments should pay closer attention and lend greater support to agriculture and food policies regarding developing nations in need.

Public institutions—foundations, agricultural departments in universities, and other national and international agricultural research organizations—should have this same sort of realization when planning their agendas and areas of focus. In addition to the national security issues, they must recognize the problem of continuing hunger and malnutrition as an important public welfare problem.

From a technological standpoint, private companies are in a position of power because they possess the scientific knowledge and capabilities to produce GM crop seeds that could have significant benefit worldwide. However, unless companies use that power for global good, their products (i.e., GM crop seeds) may continue to be stigmatized in many parts of the world, with serious market implications. Therefore, private companies should use their technological know how to focus on the needs of developing-world farmers and should partner with public institutions to benefit from a mutual sharing of resources.

Nongovernmental organizations should strive to present more balanced perspectives on the GM crop issue, keeping in mind their increased level of influence (and corresponding responsibility) in re cent years regarding policy•decisions on adoption of new technologies. NGOs that support the GM crop movement should make it clear that not all the potential risks of agricultural biotechnology have been researched.

NGOs that are against GM crops should not mislead the public about any risks that have already been proven to be insignificant, nor decline to spread the message about potential benefits from GM crops. All NGOs should help to communicate the message that the risks associated with planting certain types of GM crops in specific locales worldwide should be carefully considered.

The challenges discussed in this chapter are interrelated. Revised regulations on genetically modified crops must accompany widespread collective policy efforts to revitalize agricultural development. And before developing-world farmers and consumers can benefit from GM crops or any other type of enhanced crop breeding, the technologies must be affordable and farmers must understand how to use them.

The GM crop movement must overcome an intertwined collection of challenges before it can have an impact beyond those regions of the world that already produce excesses of food. If the GM crop movement can overcome these challenges, while proving itself to be acceptably free of adverse health and environmental impacts, it has the potential to provide benefits to farmers and consumers around the globe in previously inconceivable ways, while mitigating the need to use potentially harmful chemicals or scarce water supplies for agriculture. It can then, indeed, become a true "Gene Revolution."

8

Agriculture Biotechnology in 21st Century

FOOD SECURITY

In a world where population growth is outstripping food supply agricultural- and especially plant-biotechnology, needs to be swiftly implemented in all walks of life. The world population is expected to reach 7 billion within 25 years, over 10 billion in the year 2050, while agricultural production is growing at the slower rate of about 1.8 percentage annually. All human beings depend on agriculture that produces food of the appropriate quality at the required quantities.

Since this can no longer be achieved by traditional methods alone, *breeding through plant biotechnology is a necessity.* In the long run the massive and immediate implementation of plant and agricultural biotechnology is more highly crucial than that of medical biotechnology, since more people worldwide die from famine and diseases related to malnutrition than from "modern", western diseases.

With "Disaster Management" becoming a central issue in modern sociology research and curricula, *food and food product shortages are the ultimate disaster*. However, unlike most natural disasters, this is one that we can prepare for and even prevent.

Domestication of plants and animals found in the wild, combined with gradual long-term changes in their qualitative and quantitative traits, were the first attributes of agriculture.

Domestication, followed by food storage, coincided with the growth of microorganisms. Thus was born classical food fermentation, the earliest known application of biotechnology for the generation of food products. This traditional agriculture now faces several serious limitations:

1. Market limitations: The world is becoming a global village whose free-market rules negate the effectiveness of local pricing policies, and where a dictate of international trade and policies exists. This has greatly affected future developments in agriculture, still the world's largest business.
2. Limitations of natural resources: Global climatic changes (resulting mainly in desertification and salinization), industrialisation and urbanisation, have reduced land and water availability and caused alarming deterioration of soil, water and air quality.
3. Inherent biological genetic limitations: Although previously highly efficient, the release of new improved genotypes by classical breeding is now too slow to cope with the demands, and is considerably limited by the lack of appropriate "natural" genes that can be introgressed by traditional genetic crosses.

Only two major potential solutions seems to exist for increasing food supply and agricultural commodities, in addition to continuously improving agricultural practices, despite the aforementioned limitations:

1. A search for alternative food sources (e.g., marine or extraterrestrial products),
2. Enhanced efficient plant breeding.

COMBINING BIOTECHNOLOGY WITH CLASSICAL PHYSIOLOGY AND BREEDING

Food production, for both quantity and quality, as well as for new plant commodities and products, in developed and developing countries around the globe, cannot rely solely on classical agriculture. Human survival, vis-à-vis a continuous increase in agricultural productivity, depends on the effective *merging of classical breeding with modern plant biotechnology* and the novel tools it provides.

The "green revolution", for example, increased wheat production 10-fold in India and several other countries in South East Asia, thereby feeding three times as many people. However, this revolution has already been exploited to its limits, and alternative solutions are required to breed improved crops.

Now biotechnology, integrated with classical breeding, is on the verge of creating the "evergreen revolution". The potential to improve plant and animal productivity and their proper use in agriculture relies largely on newly developed *DNA biotechnology and molecular markers.*

These techniques enable the selection of successful genotypes, better isolation and cloning of favourable traits, and the creation of transgenic organisms of importance to agriculture. Together, these generic techniques are both an extension and an integral part of classical breeding, contributing successfully to shortening breeding and selection cycles.

The new plant biotechnology implies the use of recombinant DNA techniques and in vitro cell biology in three major areas:

1. As an aid to classical breeding: This includes the ongoing genome mapping projects, e.g., in Arabidopsis, rice, maize and tomato, combined with the recent activities in functional genomics, proteomics and bioinformatics, and DNA marker-assisted selection. The combined use of these techniques will soon shorten the time required for "classical" breeding and selection cycles.
2. Generation of engineered (transgenic) organisms: In view of the inherent limitations of introgressing new genes by traditional genetic crosses (i.e., lack of appropriate desired genes and crossing barriers), the efficient engineering of plants has already resulted in

improved field-grown transgenic plants in several important crops. The result of this impressive development, which began only 18 years ago, have made possible the direct insertion and integration of genes isolated from several organisms, and the creation of novel, and otherwise impossible genetic recombination.

3. Integration of microorganisms into plant production systems: The biotechnological development of new symbiotic, antibiotic, and antagonistic relationships between plants and microorganisms (fungi, bacteria and insects) using, among other techniques, engineered plants and microorganisms, creates new possibilities. Some of these include biological control of pests, biofertilization and plant growth stimulation, and bio- and phytoremediation.

During the last two decades, these new biotechnologies have been adapted to agricultural practices and have opened *new vistas for plant utilisation*. This will continue and intensify in the next decade. Plant biotechnology — especially in vitro regeneration and cell biology, DNA manipulation and genetic modification of biochemical pathways — is changing the plant scene in three major areas:

1. Growth and development control (vegetative, generative and propagation),
2. Protecting plants against the ever-increasing threats of abiotic and biotic stress,
3. Expanding the horizons by producing specialty foods, biochemicals and pharmaceuticals. These areas were extensively discussed at the 9th international congress of the IAPTC and B "Plant Biotechnology and In Vitro Biology in the 21st Century", held in Jerusalem in June 1998.

Vegetative, Generative and Propagation

The better insight into the control of plant regeneration, morphogenesis and patterns of cell division achieved during the last two decade, is due to three major discoveries:

1. The *totipotency and regeneration ability of plant cells and tissues*, as revealed by cell culture and micropropagation,
2. The elucidation of *genes responsible for hormone production and activation* in plants,
3. Active research into the mechanisms and molecular control of the *cell cycle and signal transduction pathways*, in part adopted from previous studies with animal cells, in part unique for plants. These have enabled both the *control and biotechnological manipulation* of vegetative growth, generative patterns (e.g., of flowers and seeds) and of micropropagation.

Vegetative Growth

Morphogenetic control mechanisms are still extremely obscure, but the advent of molecular hormone and cell-cycle research is sure to lead to a better understanding of vegetative growth patterns.

Thus, the possibility of biotechnologically manipulating *plant growth rate and architecture* can become a reality. For example, potential consequences of controlled auxin overproduction/availability include: adventitious root formation of importance to propagation, cell and organ elongation for biomass production, increased apical dominance of importance to timber production, etc.

Controlled cytokinin overproduction/availability can result, among other things, in enhanced bud break — which is of great importance to plant architecture, branching and compactness — a desired characteristic for some ornamentals, and delayed leaf and plant senescence.

No less important, in this respect, is the potential - as yet not practical - of affecting the orientation and rate of cell division, cell elongation and tissue longevity, by interfering with the cytoskeleton and cell cycle, the synthesis of cellulose and other cell components, and programmed cell death, respectively. A few of these possibilities have already been realised.

Generative Development

Flowers, fruits and seeds are extremely important for agriculture. Hence, biotechnological research and development aims to interfere with and control their development and characteristics, and some of the many related studies have already produced practical applications. The major targets in *flower development* are colour, scent and senescence.

Strategies for the molecular breeding of flower colour and scent include over- and underexpression of colour (anthocyanins and carotenoids) and scent (volatiles) compounds, with respect to their biosynthesis, cellular transport and targeting. Important targets for controlling *fruit development* include growth, ripening and senescence (as for vegetative growth), colour and scent (similar to flowers) and, in addition, flavour – particularly metabolic control of sugar, acid and other flavour components.

Of great importance to fruits are biotechnological strategies for the production of seedless fruits via parthenocarpy (overproduction of auxin), pollen destruction (no fertilisation), or arrest of embryo development. The manipulation of *seed development* using biotechnological strategies is especially critical, since the seed industry (together with vegetative propagation material) constitutes the germplasm of the future for any type of plant production system.

Seeds and vegetative propagules are, practically speaking, packages of genes that form the basis of all advanced and economically viable agricultural industries, both national and private. Biotechniques and molecular strategies are now available for the major seed-based operations: hybrid seed production, generation of artificial seeds (coated somatic embryos), and for the establishment of germplasm banks that may solve some of the biodiversity issues.

Micropropagation

Micropropagation is used routinely to generate a large number of high-quality clonal agricultural plants, including ornamental and vegetable species, and in some cases also plantation crops, fruits and vegetable species. Micropropagation has significant advantages over traditional clonal propagation techniques.

These include the potential of combining rapid large-scale propagation of new genotypes, the use of small amounts of original germplasm (particularly at the early breeding and/or transformation stage, when only a few plants are available), and the generation of pathogen-free propagules.

This impressive application of the *principles of plant cell division and regeneration to practical plant propagation* is the result of continuous tedious studies in hundreds of labouratories worldwide, many of them in developing countries, on the standardisation of explant sources, media composition and physical state, environmental conditions and acclimatisation of in vitro plants. Particularly noteworthy are the many recent studies on the molecular of organogenesis and somatic embryogenesis.

However, further practical applications of micropropagation, which is also commercially viable, depends on *reducing the production costs* such that it can compete with seed production or traditional vegetative propagation methods (e.g., cuttings, tubers and bulbs, grafting).

Techniques that have the potential to further increase the efficiency of micropropagation, but still await further improvements, include: simplified large-scale bioreactors, cheaper automatization facilities, efficient somatic embryogenesis and synthetic seed production, greater utilisation of the autotrophic growth potential of cultures, and good repeatability and quality assurance of the micropropagated plants.

Abiotic and Biotic Stress Tolerance

The application of molecular genetics and plant transformation to the *diagnosis and control of plant pests* has become one of the major practical success stories of plant biotechnology in the past decade. The availability of dozens of transgenic crop plants which are resistant to a range of insects, viruses and herbicides, as well as to several phytopathogenic fungi and nematodes has been validated under both field and laboratory conditions, and is of great economic importance.

Moreover, applying the principles of engineering plants for resistance to these pests to other plants of agricultural importance is now considered routine, although in practice still labourious, especially for new genotypes. Apart from a wider application to additional plants, the real *challenges lying ahead* include:

1. Improved expression of the target genes in the plants, especially their spatial and temporal control,
2. The use of wide-spectrum and alternative target genes to circumvent the problem of pest resistance,
3. Intensified integration of biological control via the use of selected and engineered microorganisms with a biocontrol potential.

While plant biotechnology has been applied successfully to fighting a large number of pests, this is not yet the case for abiotic stress conditions such as drought, salinity, extreme temperatures, chemical toxicity and oxidative stress. *Drought and salinization are the most common natural causes of lack of food and famine* in arid and semiarid regions, and the most serious environmental threats to agriculture in many parts of the world.

Desertification, resulting from overexploitation by the local inhabitants, is often aggravated by regional climatic changes, and results in increased soil erosion and a decrease in land and agricultural productivity. It is estimated that increased salinization of arable land will have devastating global effects, resulting in 30 percentage land loss within the next 25 years, and up to 50 percentage in the year 2050.

Although more difficult to control and engineer than the usually monogenic traits of resistance to biotic pests and herbicides, the genetically complex response to abiotic stress is globally and regionally far more important. Therefore, *breeding for plant tolerance to drought and salinity stress* should be given a high research priority in all future agbiotech programmes.

Strategies for the manipulation of *osmotic stress tolerance* in plants might include: expression of osmoprotectants and compatible solutes, ion and water transport and channels, expression of water-binding and membrane-associated dehydrins and other proteins, transcription factors and DNA-binding proteins, etc.

Also of specific interest are the intervening stages of stress perception, signal transduction (ABA and others), and protein modification. The discovery of new stress-related genes and the design of stress-specific promoters are equally important.

Food, Biochemicals and Pharmaceuticals

Traditionally, agriculture was targeted to improving the production of plant-derived food, in terms of both quantity and quality. This was also the initial primary target of plant biotechnology. The second phase of plant biotechnology is now gradually being implemented: a shift from the production of low-priced food and bulk commodities to high-priced, specialised plant-derived products.

This includes two major categories of biomaterials:

1. Direct improvement and modification of specialised constituents of plant origin,
2. The manufacture in plants of non-plant compounds. Biotechniques, mostly based on the *engineering of metabolic pathways*, are now available for modifying many plant constituents that are *used in the food, chemical and energy industries*. This includes many "primary" metabolites: carbohydrates (starch synthesis, yield and allocation, production of high-amylose or high-amylopectin starch, increased sucrose synthesis for the sugar industry, fructan production, etc.), proteins (improvement of amino acid composition and protein content), oils and fats (ratio

of saturated to nonsaturated fatty acids, increased content of specific valuable fatty acids like erucic acid, ricinoleic acid and others).

Many other plant constituents are either minor or non-food components, but have specific high- value applications, such as specific fatty acids as an alternative energy source, polysaccharides with heat hysteresis properties and for bioaffinity purification, temperature and salt-resistant enzymes for the food industry, etc.

Moreover, the use of plants as *"bioreactors" for the production of "foreign", non-plant compounds* is gaining momentum and may eventually lead to alternative types of agriculture.

This includes, for example, production of bioactive peptides, vaccines, antibodies and a range of enzymes — mostly for the pharmaceutical industry. For the chemical industry, plants can be used to produce, e.g., polyhydroxybutyrate for the production of biodegradable thermoplastics, and cyclodextrins, which form inclusion complexes with hydrophobic substances.

Achievements today in plant biotechnology have already surpassed all previous expectations, and the future is even more promising. The full realisation and impact of the aforementioned developments depend, however, not only on continued successful and innovative research and development activities, but also on a favourable regulatory climate and public acceptance.

About 12 percentage of the world's land surface is used to grow crops, and the agricultural area required to support food production — 0.44 ha / capita in 1961 — will probably have been reduced to 0.15 ha / capita in 2050.

The intensification of agriculture with its aforementioned limitations thus requires enhanced and more efficient plant breeding and the release of economical, high-return and patentable plant-derived products. This cannot be achieved without supporting advanced research and development in biochemistry, physiology, genomics and biotechnology of agricultural plants.

Plant scientists now have a central role in society, not unlike their place 300 years ago when Jonathan Swift stated: "Whoever could make two ears of corn or two blades of grass to grow upon a spot of ground where only one grew before, would deserve better of mankind, and do more essential service to his country, than the whole race of politicians put together."

9

Importance of Agriculture Biotechnology

GENETICALLY MODIFIED FOODS

It is our quest as technology education teachers to prepare our students to become technologically literate and productive members of our society—a daunting task in a society growing ever dependent on our technological development. The challenge becomes formidable in our attempt to not only understand the complexities and impacts these technologies might have on our society, and indeed on the whole human race, but at the same time impart an understanding to our students so that they have the opportunity to become technologically literate with the capability to make sound rational decisions in the use and deployment of these technologies for the betterment of our planet and all its inhabitants.

One area of technological development that cannot be ignored in our classrooms is that of genetically modified foods. The question is not whether we should modify the genetics of our crop plants, which has been going on since plants were first domesticated—the real questions should focus on how new genetically engineered modifications will affect our society and how genetically modified crops and plants will alter our environment. The term GM food is used in reference to crop plants modified for human and animal consumption using the latest molecular biological techniques. It is the purposeful addition of a foreign gene or genes to the genome of an organism.

A gene holds the information that will give the organism trait. Genetic engineering physically removes the DNA from one organism and transfers the gene(s) for one or a few traits to another. Since cross breeding between two sexually compatible plants is not necessary, the sexual barrier between species is overcome. These modern techniques now enable scientists to enhance desired traits such as increased resistance to herbicides or improved nutritional content, in ways they could not before, and with greater ease and precision.

Genetic engineering is not bound by the limitations of traditional plant breeding—not only can genes be transferred from one plant to another, but genes from non-plant organisms can also be used. The best-known example of this is the use of Bt genes in corn and other crops. Bacillus thuringiensis is a naturally occurring bacterium that produces crystal proteins that are lethal to insect larvae. These proteins have been transferred into corn and cotton so that these plants can create their own pesticides.

Through genetic manipulation, crop plants are being modified for pest resistance, herbicide tolerance, disease resistance, cold tolerance, drought tolerance, salinity tolerance, and increased nutrition. Researchers are also working to develop edible vaccines in tomatoes and potatoes. Plants such as poplar trees have been genetically engineered to clean up heavy metal pollution from contaminated soil.

Genetically Modified (GM) foods are produced from genetically modified organisms (GMO) which have had their genome altered through genetic engineering techniques. The general principle of producing a GMO is to insert DNA that has been taken from another organism and modified in the laboratory into an organism's genome to produce both new and useful traits and phenotypes. Typically this is done using DNA from certain types of bacteria.

GM Foods have been available since the 1990s, with the principal ones being derived from plants; soybean, maize, canola and cotton seed oil. The first commercially grown genetically modified food crop was the Flavr Savr tomato which was made more resistant to rotting by Californian company Calgene. Calgene was allowed to release it into the market in 1994 without any special labeling, where it was welcomed by consumers who purchased the fruit at two to five times the price of standard tomatoes.

However, production problems and competition from a conventionally bred, longer shelf-life variety prevented the product from becoming profitable. A variant of the Flavr Savr was used by Zeneca to produce tomato paste which was sold in Europe during the summer of 1996. It's labeling and pricing were designed as a marketing experiment which proved that, at the time, European consumers would accept genetically engineered foods. This attitude would be drastically changed after outbreaks of Mad Cow Disease weakened consumer trust in government regulators, and protesters rallied against the introduction of Monsanto's "Roundup-Ready" soybeans.

The next GM crops included insect-protected cotton and herbicide-tolerant soybeans both of which were commercially released in 1996. These crops have been widely adopted in the United States. They have also been extensively planted in several other countries (Argentina, Brazil, South Africa, India, and China) where agriculture is a major part of the total economy. Other GM crops include insect-protected maize and herbicide-tolerant maize, cotton, and rapeseed varieties.

CROPS UNDER DEVELOPMENT

The following GM crops are in development.

- Sweet potato resistant to the feathery mottle virus
- Further development of golden rice to increase levels or bioavailability of iron, zinc, vitamin A, vitamin E and improve the quality of proteins
- Maize with increased levels of the amino acid lysine and protein for animal feeds
- A variety of plants able to better tolerate non-biological stresses which are commonly encountered in a normal

growing season, such as water and nitrogen limitation, or survive extreme growing conditions, such as high-salinity, drought, acidic soils, or hot weather. Such traits can provide more reliable crop performance over an extended period of cultivation.

- Transgenic rice has been developed by a Californian company to improve oral rehydration therapy for diarrhea. In sub-Saharan Africa and parts of Latin America and Asia, diarrhea is the second highest infectious killer of children under the age of five, accounting for some two million deaths a year. Recent 2005-6 trials in a Peruvian Hospital have demonstrated that specialized milk proteins lactoferrin and lysozyme made in transgenic rice plants improve the effectiveness of oral rehydration solution used to treat diarrhea.

Abundance of GM Crops

Between 1996 and 2005, the total surface area of land cultivated with GMOs had increased by a factor of 50, from 17,000 km^2 (4.2 million acres) to 900,000 km^2 (222 million acres), of which 55% were in the United States.

Although most GM crops are grown in North America, in recent years there has been rapid growth in the area sown in developing countries. For instance in 2005 the largest increase in crop area planted to GM crops (soybeans) was in Brazil (94,000 km^2 in 2005 versus 50,000 km^2 in 2004.) There has also been rapid and continuing expansion of GM cotton varieties in India since 2002. (Cotton is a major source of vegetable cooking oil and animal feed.) It is predicted that in 2006/7 32,000 km^2 of GM cotton will be harvested in India (up more than 100% from the previous season).

Indian national average cotton yields have been boosted to close 50% above the long term average yield during this period. The publicity given to transgenic trait Bt insect resistance has encouraged the adoption of better performing hybrid cotton varieties, and the Bt trait has substantially reduced losses to insect predation. Economic and environmental benefits of GM cotton in India to the individual farmer have been documented.

In 2003, countries that grew 99% of the global transgenic crops were the United States (63%), Argentina (21%), Canada (6%), Brazil (4%), China (4%), and South Africa (1%). The Grocery Manufacturers of America estimate that 75% of all processed foods in the U.S. contain a GM ingredient. In particular, Bt corn, which produces the pesticide within the plant itself is widely grown, as are soybeans genetically designed to tolerate glyphosate herbicides. These constitute "input-traits" that financially benefit the producers, have indirect environmental benefits and marginal cost benefits to consumers.

In the US, by 2006 89% of the planted area of soybeans, 83% of cotton, and 61% maize was genetically modified varieties. Genetically modified soybeans carried herbicide tolerant traits only, but maize and cotton carried both herbicide tolerance and insect protection traits (the latter largely the *Bacillus thuringiensus* Bt insecticidal protein). In the period 2002 to 2006, there were significant increases in the area planted to Bt protected cotton and maize, and herbicide tolerant maize also increased in sown area.

Future Developments

Future envisaged applications of GMOs are diverse and include drugs in food, bananas that produce human vaccines against infectious diseases such as Hepatitis B, metabolically engineered fish that mature more quickly, fruit and nut trees that yield years earlier, and plants that produce new plastics with unique properties. While their practicality or efficacy in commercial production has yet to be fully tested, the next decade may see exponential increases in GM product development as researchers gain increasing access to genomic resources that are applicable to organisms beyond the scope of individual projects.

Safety testing of these products will also at the same time be necessary to ensure that the perceived benefits will indeed outweigh the perceived and hidden costs of development. Controversies surrounding GM foods and crops commonly focus on human and environmental safety, labeling and consumer choice, intellectual property rights, ethics, food security, poverty reduction, and environmental conservation. See also: GM food controversy

Safety Testing

In the USA regulation of a genetically modified food is determined by the objective characteristics of the food and the intended use of the food, irrespective of the way it was developed. FDA policy states that a formal pre-market review by the FDA is to be taken when the objective characteristics of any substance added to the food raises safety issues.

Prior to marketing a new GM food product, manufacturers are required to submit documentation to the FDA to demonstrate its safety and then await approval before selling it to consumers.

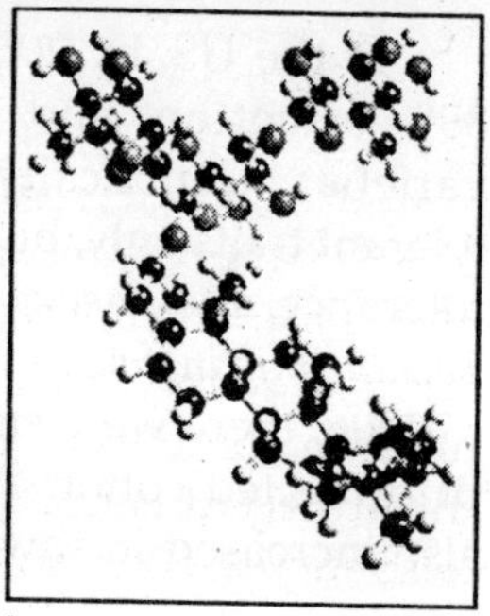

Fig. Solanine

The context for assessing safety of novel foods is the fact that existing foods often contain toxic components but are still able to be consumed safely. For instance, potatoes and tomatoes can contain toxic levels of solanine and alpha-tomatine alkaloids respectively, and this situation is recognised in the concept of "Substantial Equivalence" that was developed by the OECD in 1993 as a criterion for identifying whether a novel food is at least as safe as the equivalent existing food. The US FDA takes a safety assessment approach that is consistent with this OECD concept in their regulation of novel foods (including those made by recombinant DNA methods). This policy is outlined in an FDA statement.

Critics of GM food believe this regulatory model fails to sufficiently protect consumers and claim that the FDA is subject to pressure and influence by industry. One concern voiced is that a novel crop may have unintended changes created during the insertion of new genetic material. On the other hand, plant scientists, backed by results of modern comprehensive profiling of crop composition, point out that crops modified using GM techniques are less likely to have unintended changes than are conventionally bred crops.

Intellectual Property

Enforcement of Patents on genetically modified plants is often contentious, especially because of the occurrence of Gene flow. In 1998, 95-98% of about 10 km^2 planted with canola by Canadian farmer Percy Schmeiser were found to contain Monsanto's patented Roundup Ready gene although Schmeiser had never purchased seed from Monsanto. The initial source of the plants was undetermined, and could have been through either gene flow or intentional theft. However, the overwhelming predominance of the trait implied that Schmeiser must have intentionally selected for it.

The court determined that Schmeiser had saved seed from areas on and adjacent to his property where Roundup had been sprayed, such as ditches and near power poles. Although unable to prove direct theft, Monsanto sued Schmeiser for piracy since he knowingly grew Roundup Ready plants without paying royalties.

The case made it to the Canadian Supreme Court, which in 2004 ruled 5 to 4 in Monsanto's favour. The dissenting judges focused primarily on the fact that Monsanto's patents covered only the gene itself and glyphosate resistant *cells*, and failed to cover transgenic plants in their entirety.

In response to criticism, Monsanto Canada's director of public affairs stated that "It is not, nor has it ever been Monsanto Canada's policy to enforce its patent on Roundup Ready crops when they are present on a farmer's field by accident...Only when there has been a knowing and deliberate violation of its patent rights will Monsanto act." Currently Percy Schmeiser spends a large amount of his time traveling and speaking about how Monsanto ruined his career as a farmer. He also talks about the possible harms of genetically modified and why others in addition to himself should be protesting it.

Benefits and Controversies

Some argue that there is more than enough food in the world and that the hunger crisis is caused by problems in food distribution and politics, not production, so people should not be offered food that may carry some degree of risk.

Others oppose genetic engineering on the grounds that genetic modifications might have unforeseen consequences, both in the initially modified organisms and their environments. For example, certain strains of maize have been developed that are toxic to plant eating insects. It has been alleged those strains cross-pollinated with other varieties of wild and domestic maize and passed on these genes with a putative impact on Maize biodiversity.

Subsequent to the publication of these results, several scientists pointed out that the conclusions were based on experiments with design flaws. It is well known that the results from the Polymerase Chain Reaction method of analysing DNA can often be confounded by sample contamination and experimental artifacts. Appropriate controls can be included in experiments to eliminate these as a possible explanation of the results - however these controls were not included in the methods used by Quist and Chapela.

After this criticism *Nature,* the scientific journal where this data was originally published concluded that "the evidence available is not sufficient to justify the publication of the original paper". More recent attempts to replicate the original studies have concluded that genetically modified corn is absent from southern Mexico in 2003 and 2004. Also in dispute is the impact on biodiversity of the introgression of transgenes into wild populations. Unless a transgene offers a massive selective advantage in a wild population, a transgene that enters such a population will be maintained at a low gene frequency. In such situations it can be argued that such an introgression actually *increases* biodiversity rather than lowers it.

Activists opposed to genetic engineering say that with current recombinant technology there is no way to ensure that genetically modified organisms will remain under control, and the use of this technology outside secure laboratory environments carries potentially unacceptable risks to both farmed and wild ecosystems.

Potential impact on biodiversity may occur if herbicide-tolerant crops are sprayed with herbicide to the extent that no wild plants ('weeds') are able to survive. Plants toxic to insects may mean insect-free crops. This could result in declines in other wildlife (e.g. birds) which feed on weed seeds and/or insects for food resources. The recent (2003) farm scale studies in the UK found this to be the case with GM sugar beet and GM rapeseed, but not with GM maize (though in the last instance, the non-GM comparison maize crop had also been treated with environmentally-damaging pesticides subsequently (2004) withdrawn from use in the EU).

Although some scientists have claimed that selective breeding is a form of genetic engineering, (e.g., maize was modified from teosinte, dogs have evolved with human intervention over the course of tens of thousands of years from wolves), others assert that modern transgenesis-based genetic engineering is capable of delivering changes faster than, and sometimes of different types from, traditional breeding methods.

Proponents of current genetic techniques as applied to food plants cite the benefits that the technology can have, for example, in the harsh agricultural conditions of Africa. They say that with modifications, existing crops would be able to thrive under the relatively hostile conditions providing much needed food to their people. Proponents also cite golden rice and golden rice 2, genetically engineered rice varieties (still under development) that contain elevated vitamin A levels. There is hope that this rice may alleviate vitamin A deficiency that contributes to the death of millions and permanent blindness of 500,000 annually.

Proponents say that genetically-engineered crops are not significantly different from those modified by nature or humans in the past, and are as safe or even safer than such methods. There is gene transfer between unicellular eukaryotes and prokaryotes. There have been no known genetic catastrophes as a result of this. They argue that animal husbandry and crop breeding are also forms of genetic engineering that use artificial selection instead of modern genetic modification techniques. It is politics, they argue, not economics or science, that causes their work to be closely investigated, and for different standards to apply to it than those applied to other forms of agricultural technology.

Proponents also note that species or genera barriers have been crossed in nature in the past. An oft-cited example is today's modern red wheat variety, which is the result of two natural crossings made long ago. It is made up of three groups of seven chromosomes. Each of those three groups came from a different wild wheat grass. First, a cross between two of the grasses occurred, creating the durum wheats, which were the commercial grains of the first civilizations up through the Roman Republic. Then a cross occurred between that 14-chromosome durum wheat and another wild grass to create what became modern red wheat at the time of the Roman Empire.

Economic and Political Effects

- Many opponents of current genetic engineering believe the increasing use of GM in major crops has caused a power shift in agriculture towards Biotechnology companies, which are gaining excessive control over the production chain of crops and food, and over the farmers that use their products, as well.
- Many proponents of some current genetic engineering techniques believe it will lower pesticide usage and has brought higher yields and profitability to many farmers, including those in developing nations. A few genetic engineering licenses allow farmers in less

economically developed countries to save seeds for next year's planting.

- In August 2002, Zambia cut off the flow of Genetically Modified Food (mostly maize) from UN's World Food Programme. Although there were claims that this left a famine-stricken population without food aid, the U.N. programme succeeded in replacing the rejected grain with other sources, including some foods purchased locally with European cash donations. In rejecting the maize, Zambians cited the "Precautionary Principle" and also the desire to protect future possibilities of grain exports to Europe.
- In December 2005 the Zambian government changed its mind in the face of further famine and allowed the importation of GM maize. However, the Zambian Minister for Agriculture Mundia Sikatana has insisted that the ban on genetically modified maize remains, saying "We do not want GM (genetically modified) foods and our hope is that all of us can continue to produce non-GM foods."
- In April 2004 Hugo Chavez announced a total ban on genetically modified seeds in Venezuela.
- In January 2005, the Hungarian government announced a ban on importing and planting of genetic modified maize seeds, although these were agreed authorized by the EU.
- On August 18, 2006, American exports of rice to Europe were interrupted when much of the U.S. crop was confirmed to be contaminated with unapproved engineered genes, possibly due to accidental cross-pollination with conventional crops. The U.S. government has since declared the rice safe for human consumption, and exports to some countries have since resumed.

Compared with traditional cross breeding techniques where hundreds and thousands of genes are transferred indiscriminately, using genetic engineering techniques to develop GM crops with the introduction of one or a few genes results in subtle and less disruptive changes that are relatively specific and predictable.

Direct gene transfer is still a relatively new procedure and can be viewed as a logical extension to the methods that have been used for thousands of years to modify our crop plants. Even so, many questions have arisen concerning the safety of these foods and their potential impact on the environment.

SAFETY OF BIOTECH FOODS

The first genetically engineered whole product available to the marketplace was the Flavr Savr. tomato, introduced to the public in May 1994 without much fanfare. The Flavr Savr. tomato was genetically engineered by Calgene, Inc., to stay ripe without rotting for up to ten days, allowing plenty of time for the vine-ripened tomato to be picked ripe, shipped, and sold. Calgene, although not required to, sought approval by the FDA to assure consumers that their product was safe for consumption.

Since the approval of the Flavr Savr. tomato, every developer of biotech foods has, without exception, consulted with the FDA on a voluntary basis to preemptively address any safety or nutrition related concerns before the product reaches the marketplace. In the eight years since the introduction of the Flavr Savr. tomato, there hasn't been a single documented case of an illness caused by biotech foods.

In the U.S., the safety of biotech foods is overseen by three separate agencies:

- The Food and Drug Administration (FDA) assesses the safety of all foods and animal feeds, including those produced through plant biotechnology.
- The Department of Agriculture (USDA), through its Animal and Plant Health Inspection Service, oversees field-testing of biotech seeds and plants to make sure their release causes no harm to the environment, especially native plants.

- The Environmental Protection Agency (EPA) evaluates biotech plants' environmental safety such as their pesticide properties, possible effect on wildlife, and how these plants break down in the environment.

Issues of concern have also been raised with regard to biotech foods and the environment. Cornell University researchers found, while conducting laboratory studies, that monarch larvae could be harmed or killed if they ate large amounts of Bt corn pollen. When Cornell University researchers conducted their 1999 study they did not attempt to duplicate real-world environmental conditions.

They used only a small number of caterpillars and gave them no choice but to eat leaves coated with a thick layer of Bt corn pollen. Subsequent field studies conducted by the USDA's Agriculture Research Service (ARS) demonstrated that Bt corn does not harm the monarch butterfly. Extensive studies from around the world show that biotech crops are helping to preserve and protect the environment.

Researchers from Washington State University, the University of Illinois, Clemson University, and the National Centre for Food and Agricultural Policy (NCFAP) found that through the use of biotech crops, the following is possible:

- Valuable soil erosion can be prevented using environmentally friendly no-till farming practices when planting biotech soybeans and cotton.
- The ability to use more benign herbicides that rapidly dissipate in soil and water improves water quality.
- Air quality is improved when no-till farming practices are used, significantly reducing the release of greenhouse gasses.

There are also concerns that eating GM foods could lead to an increase of unintentional allergic reactions. Extensive testing is part of the approval process to make sure that proteins introduced to crops to improve their characteristics do not cause new allergies. Scientists believe that biotechnology is a promising tool for removing allergens from foods, giving many allergy-prone people a wider choice of safe foods to eat.

Even before they reach a farmer's field, biotech corn, soybeans, and other genetically enhanced foods undergo years of review by researchers, university scientists, farmers, and other government agencies in addition to the FDA. The results of these tests have all produced the same conclusions. Biotech crops are safe to eat. A report issued by the National Academy of Sciences, an independent group of scientists and scholars, confirmed that all approved biotech products are as safe as their conventional counterparts for human and animal consumption.

Benefits of GM Foods

With a continuously expanding world population that already exceeds six billion people, ensuring an adequate food supply is going to become a major challenge. GM foods can potentially help to meet this challenge in a number of ways:

- Pest resistant crops—world crop productivity could increase by as much as 25 percent through the use of biotechnology to grow plants that resist pests and diseases. Growing GM foods such as Bt corn can help eliminate the application of chemical pesticides, reducing the run-off of agriculture wastes that can poison valuable water sources and cause harm to the environment.
- Disease resistant crops—scientists are creating plants that have the ability to fight against many viruses, fungi, and bacteria that are harmful to them.
- Cold tolerance—by introducing an antifreeze gene from cold water fish, plants are being developed that are able to withstand cold temperatures that would normally kill them.
- Drought tolerance/salinity tolerance—plants created to withstand periods of drought or high salt content in their soil and groundwater will help to grow crops in land formerly unsuitable for growing.

- Pharmaceuticals—vaccines introduced into tomatoes and potatoes are much easier to ship, store, and administer than traditional vaccines that have to be injected, a major benefit for developing third-world countries.

International food and agricultural experts and policy makers—including the U.N. Food and Agricultural Organization and the World Health Organization—call plant biotechnology a critical tool to help feed a growing population in the twenty-first century. Using existing farmland, world crop productivity could increase by as much as 25 percent through the use of GM crops engineered to resist pests and diseases, tolerate harsh growing conditions, and delay ripening to reduce spoilage.

A study conducted by the National Centre for Food and Agricultural Policy (NCFAP), released in June 2002, found that six biotech crops planted in the U.S.—soybeans, corn, cotton, papaya, squash, and canola—produced an additional four billion pounds of food and fiber on the same acreage, improved farm income by $1.5 billion, and reduced pesticide use by 46 million pounds.

While some nations vacillate about the merits of genetically modified crops, other nations such as China, Africa, India, and Pacific Island nations embrace GM crops as a means to expand production of food crops in order to feed the world's fastest-growing populations. Rice is a staple crop for nearly a billion people in Asian countries where 50 percent or more of calories are from rice. Aside from providing more calories from higher yields, genetically modified rice can provide increased nutrients such as vitamin A, lysine, iron, and zinc, thus reducing illnesses and malnutrition associated with deficiencies of these nutrients.

Labeling Concerns

The labeling of GM foods and food products is a controversial issue. Food industries support the belief that labeling should be voluntary and influenced by consumer demands and say that requiring a label for biotech ingredients would confuse consumers more than it would inform them. Consumer interest groups are demanding mandatory labeling, contending that people have the right to know what they are eating.

The Food and Drug Administration (FDA), which oversees food safety issues, performs exhaustive safety tests on every biotech food product entering the marketplace and has concluded that special labeling is required only when the new food product is significantly different from its conventional counterpart. The American Medical Association (AMA) has also stated that, "there is no scientific justification for special labeling of genetically modified foods, as a class".

Special labeling requirements for GM foods would dramatically increase the costs of all foods—costs that would be passed on to the consumer. The FDA is currently developing voluntary labeling guidelines, which would enable manufacturers to let consumers know if a food was developed using biotechnology to have a beneficial trait, or if biotech ingredients were used in making a product. Any type of food labeling should be designed to clearly convey accurate information about the product in language that everyone can understand. The biggest challenge in any GM food labeling policy would be how to educate and inform the public without causing undue alarm or fear of GM foods.

Each year the world's population increases by approximately 73 million people and is projected to reach approximately eight billion by the year 2030—with most of the increase expected to occur in developing countries. Despite efforts worldwide to combat hunger and malnutrition, there are over 800 million people worldwide who do not receive adequate food and nutrition. In order for the world to address this critical situation, the world will have to double its food production and distribution over the next 25 years.

Plant biotechnology can help to overcome the world's concern for feeding its ever-growing population safely and effectively by continuing to develop a food supply that is more abundant on less acreage, has more nutritional value, and is more environmentally conservative.

Governments around the world need to develop policies to ensure greater investment in research and at the same time develop necessary regulations and guidelines that will enable continued advances in plant biotechnology while maintaining confidence and trust that our food supply is safe for consumption and the ecosystem of our world.

Standards for Technological Literacy

Introducing students to plant biotechnology will expose students to the relatively new science of genetically engineered foods and help them to understand the concerns, controversy, and potential that this technology espouses. As more and more foods are introduced to our daily diet, it is important that students understand the methods used to develop these foods, regulations, and guidelines used to determine their safety for human consumption, and tests that are performed to ensure that our environment is not harmed.

Standards for Technological Literacy offers benchmark guides and suggestions that will help technology education teachers in their preparation to expose their students to the wonders and potential of plant biotechnology.

Genetic Engineering Activity: Regulation and Policies for Genetically Modified Crops

Objectives:

- Students will become informed about the processes used to genetically modify crops.
- Students will assess the impact of genetically modified crops on society and the environment and make intelligent decisions about their value and how they should be regulated.

As we have seen, genetic engineering has the potential to alter plant and animal life forms in ways not found in nature. For centuries, farmers have used selective breeding to improve both crops and animals that had the characteristics they wanted to bring out and strengthen. This was the only way that farmers had to develop plants and animals that were resistant to disease and tolerant of climate extremes.

Today, scientists can find individual genes that control particular characteristics, separate them out, change them, and transfer them directly into the cells of an animal, plant, bacterium, or virus. Because the DNA code is known and is common to all life, it is also possible to produce synthetic genes. This technology is called genetic modification or genetic engineering.

Genetic modification means that, for the first time, humans can make living things to our own design, without relying on nature. The implications for this new technology are vast. Although any new technology may have its risks and benefits, genetic engineering has special features that need to be addressed with wisdom and insight.

There is much controversy about genetically engineered or modified plants such as Bt corn (bacillus thuringiensis), Bt cotton, and soybeans. One side is that genetically modified crops provide benefits such as increased crop yields, reduced pesticide use, and consistent quality under varying growing conditions, and improved economic returns. Additionally, GM crops could help meet the nutritional needs of millions of people around the globe who are literally starving.

Conversely, there are risks in developing and releasing GM crops for food consumption before they are thoroughly tested. The media recently reported that a public interest group found StarLink's genetically modified protein in taco shells, sparking contentious international debate over crop biotechnology. The claim was made that the Starlink[R] contained an allergen that may affect some people. Subsequently, Starlink[R] has been banned for human consumption. Could a GM plant cross-pollinate with another plant to produce a high-risk offspring? These issues and others present real challenges to companies developing GM plants as well as to governmental and regulatory agencies and consumers here and abroad.

As a student of technology, you are aware of the assessment process used to evaluate existing and emerging technologies. Given the fact that GM crops are an invention rather than a product of nature and natural selection, it would be appropriate to assess the benefits of GM crops and propose a set of policies that could be used to regulate them. It is important to note that care must be used in establishing policies that will encourage the development of new GM crops, yet protect the public and environment. Policies that are too restrictive will inhibit innovation and development and become a barrier to the introduction of GM crops. Policies that are ineffective may result in GM crops that could create irreversible impacts on the ecological environment.

The following are issues that can provide a starting point for developing a set of policies regarding genetically engineered crops.

- Benefit to humans—nutritional and economic
- Human consumption risks—allergic and intestinal reactions
- Intellectual property-patents, ownership, licensing, and royalties
- Environmental risks—effects on other plants and insects
- Ecological risks
- Economic impacts
- Safety and testing
- Control and monitoring the distribution of GM seed and produce
- Liability, compensation, and risk management
- Should genetically modified foods be labeled?

Plant Breeding

Plant breeding is the purposeful manipulation of plant species in order to create desired genotypes and phenotypes for specific purposes. This manipulation involves either controlled pollination, genetic engineering, or both, followed by artificial selection of progeny. *Plant breeding* often, but not always, leads to plant domestication.

Plant breeding has been practiced for thousands of years, since near the beginning of human civilization. It is now practiced worldwide by government institutions and commercial enterprises. International development agencies believe that breeding new crops is important for ensuring food security and developing practices of sustainable agriculture through the development of crops suitable for their environment.

Domestication

Domestication of plants is an artificial selection process conducted by humans to produce plants that have fewer undesirable traits of wild plants, and which renders them dependent on artificial (usually enhanced) environments for their continued existence. The practice is estimated to date back 9,000-11,000 years. Many crops in present day cultivation are the result of domestication in ancient times, about 5,000 years ago in the Old World and 3,000 years ago in the New World. In the Neolithic period, domestication took a minimum of 1,000 years and a maximum of 7,000 years. Today, all of our principal food crops come from domesticated varieties.

Fig. Rice (Oryza sativa)

A cultivated crop species that has evolved from wild populations due to selective pressures from traditional farmers is called a landrace. Landraces, which can be the result of natural forces or domestication, are plants (or animals) that are ideally suited to a particular region or environment. Examples are the landraces of rice, *Oryza sativa* subspecies *indica,* which was developed in South Asia, and *Oryza sativa* subspecies *japonica,* which was developed in China.

Classical Plant Breeding

Classical plant breeding uses deliberate interbreeding (crossing) of closely or distantly related individuals to produce new crop varieties or lines with desirable properties. Plants are crossbred to introduce traits/genes from one variety or line into a new genetic background. For example, a mildew-resistant pea may be crossed with a high-yielding but susceptible pea, the goal of the cross being to introduce mildew resistance without losing the high-yield characteristics.

Fig. Cisgenic plants

Progeny from the cross would then be crossed with the high-yielding parent to ensure that the progeny were most like the high-yielding parent, (backcrossing). The progeny from that cross would then be tested for yield and mildew resistance and high-yielding resistant plants would be further developed. Plants may also be crossed with themselves to produce inbred varieties for breeding.

Classical breeding relies largely on homologous recombination of two genomes to generate genetic diversity. The classical plant breeder may also makes use of a number of *in vitro* techniques such as protoplast fusion, embryo rescue or mutagenesis (see below) to generate diversity and produce Transgenic plants that would not exist in nature.

The Yecoro wheat (right) cultivar is sensitive to salinity, plants resulting from a hybrid cross with cultivar W4910 (left) show greater tolerance to high salinity

Traits that breeders have tried to incorporate into crop plants in the last 100 years include:

1. Increased quality and yield of the crop
2. Increased tolerance of environmental pressures (salinity, extreme temperature, drought)
3. Resistance to viruses, fungi and bacteria

4. Increased tolerance to insect pests
5. Increased tolerance of herbicides

Intraspecific hybridization within a plant species was demonstrated by Charles Darwin and Gregor Mendel, and was further developed by geneticists and plant breeders. In the early 20th century, plant breeders realized that Mendel's findings on the non-random nature of inheritance could be applied to seedling populations produced through deliberate pollinations to predict the frequencies of different types.

In 1908, George Harrison Shull described heterosis, also known as hybrid vigour. Heterosis describes the tendency of the progeny of a specific cross to outperform both parents. The detection of the usefulness of heterosis for plant breeding has lead to the development of inbred lines that reveal a heterotic yield advantage when they are crossed. Maize was the first species where heterosis was widely used to produce hybrids.

By the 1920s, statistical methods were developed to analyze gene action and distinguish heritable variation from variation caused by environment. In 1933, another important breeding technique, cytoplasmic male sterility (CMS), developed in maize, was described by Marcus Morton Rhoades. CMS is a maternally inherited trait that makes the plant produce sterile pollen, enabling the production of hybrids and removing the need for detasseling maize plants.

These early breeding techniques resulted in large yield increase in the United States in the early 20th century. Similar yield increases were not produced elsewhere until after World War II, the Green Revolution increased crop production in the developing world in the 1960s.

Following World War II a number of techniques were developed that allowed plant breeders to hybridize distantly related species, and artificially induce genetic diversity.

When distantly related species are crossed, plant breeders make use of a number of plant tissue culture techniques to produce progeny from otherwise fruitless mating. Interspecific and intergeneric hybrids are produced from a cross of related species or genera that do not normally sexually reproduce with each other. These crosses are referred to as *Wide crosses*. For example, the cereal triticale is a wheat and rye hybrid.

The cells in the plants derived from the first generation created from the cross contained an uneven number of chromosomes and as result was sterile. The cell division inhibitor colchicine was used to double the number of chromosomes in the cell and thus allow the production of a fertile line.

Failure to produce a hybrid may be due to pre- or post-fertilization incompatibility. If fertilization is possible between two species or genera, the hybrid embryo may abort before maturation. If this does occur the embryo resulting from an interspecific or intergeneric cross can sometimes be rescued and cultured to produce a whole plant. Such a method is referred to as *Embryo Rescue*. This technique has been used to produce new rice for Africa, an interspecific cross of Asian rice *(Oryza sativa)* and African rice *(Oryza glaberrima)*.

Hybrids may also be produced by a technique called protoplast fusion. In this case protoplasts are fused, usually in an electric field. Viable recombinants can be regenerated in culture.

Chemical mutagens like EMS and DMSO, radiation and transposons are used to generate mutants with desirable traits to be bred with other cultivars. Classical plant breeders also generate genetic diversity within a species by exploiting a process called somaclonal variation, which occurs in plants produced from tissue culture, particularly plants derived from callus. Induced polyploidy, and the addition or removal of chromosomes using a technique called chromosome engineering may also be used.

When a desirable trait has been bred into a species, a number of crosses to the favoured parent are made to make the new plant as similar as the parent as possible. Returning to the example of the mildew resistant pea being crossed with a high-yielding but susceptible pea, to make the mildew resistant progeny of the cross most like the high-yielding parent, the progeny will be crossed back to that parent for several generations. This process removes most of the genetic contribution of the mildew resistant parent. Classical breeding is therefore a cyclical process.

It should be noted that with classical breeding techniques, the breeder does not know exactly what genes have been introduced to the new cultivars. Some scientists therefore argue that plants produced by classical breeding methods should undergo the same safety testing regime as genetically modified plants. There have been instances where plants bred using classical techniques have been unsuitable for human consumption, for example the poison solanine was unintentionally increased to unacceptable levels in certain varieties of potato though plant breeding. New potato varieties are often screened for solanine levels before reaching the marketplace.

MODERN PLANT BREEDING

Modern plant breeding uses techniques of molecular biology to select, or in the case of genetic modification, to insert, desirable traits into plants.

Marker-Assisted Selection

Sometimes many different genes can influence a desirable trait in plant breeding. The use of tools such as molecular markers or DNA fingerprinting can map thousands of genes. This allows plant breeders to screen large populations of plants for those that possess the trait of interest. The screening is based on the presence or absence of a certain gene as determined by laboratory procedures, rather than on the visual identification of the expressed trait in the plant.

Genetic Modification

Genetic modification of plants is achieved by adding a specific gene or genes to a plant, or by knocking out a gene with RNAi, to produce a desirable phenotype. The plants resulting from adding a gene are often referred to as transgenic plants. If for genetic modification genes of the species or of a crossable plant are used under control of their native promoter, then they are called Cisgenic plants. Genetic modification can produce a plant with the desired trait or traits faster than classical breeding because the majority of the plant's genome is not altered.

To genetically modify a plant, a genetic construct must be designed so that the gene to be added or knocked-out will be expressed by the plant. To do this, a promoter to drive transcription and a termination sequence to stop transcription of the new gene, and the gene of genes of interest must be introduced to the plant. A marker for the selection of transformed plants is also included. In the labouratory, antibiotic resistance is a commonly used marker: plants that have been successfully transformed will grow on media containing antibiotics; plants that have not been transformed will die. In some instances markers for selection are removed by backcrossing with the parent plant prior to commercial release.

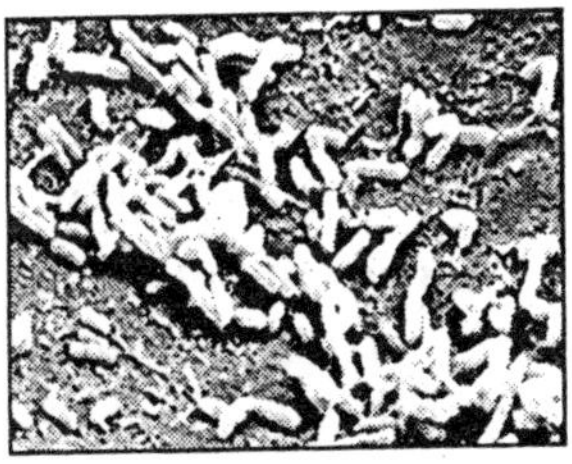

Fig. Agrobacterium tumefaciens

The construct can be inserted in the plant genome by genetic recombination using the bacteria *Agrobacterium tumefaciens* or *A. rhizogenes*, or by direct methods like the gene gun or microinjection. Using plant viruses to insert genetic constructs into plants is also a possibility, but the technique is limited by the host range of the virus. For example, Cauliflower mosaic virus (CaMV) only infects cauliflower and related species. Another limitation of viral vectors is that the virus is not usually passed on the progeny, so every plant has to be inoculated.

The majority of commercially released transgenic plants, are currently limited to plants that have introduced resistance to insect pests and herbicides. Insect resistance is achieved through incorporation of a gene from *Bacillus thuringiensis* (Bt) that encodes a protein that is toxic to some insects. For example, the cotton bollworm, a common cotton pest, feeds on Bt cotton it will ingest the toxin and die. Herbicides usually work by binding to certain plant enzymes and inhibiting their action.

The enzymes that the herbicide inhibits are known as the herbicides *target site*. Herbicide resistance can be engineered into crops by expressing a version of *target site* protein that is not inhibited by the herbicide. This is the method used to produce glyphosate resistant crop plants. Genetic modification of plants that can produce pharmaceuticals (and industrial chemicals), sometimes called *pharmacrops*, is a rather radical new area of plant breeding.

Issues and Concerns

Modern plant breeding, whether classical or through genetic engineering, comes with issues of concern, particularly with regard to food crops. The question of whether breeding can have a negative effect on nutritional value is central in this respect. Although relatively little direct research in this area has been done, there are scientific indications that, by favouring certain aspects of a plant's development, other aspects may be retarded.

A study published in the Journal of the American College of Nutrition in 2004, entitled Changes in USDA Food Composition Data for 43 Garden Crops, 1950 to 1999, compared nutritional analysis of vegetables done in 1950 and in 1999, and found substantial decreases in six of 13 nutrients measured, including 6% of protein and 38% of riboflavin. Reductions in calcium, phosphorus, iron and ascorbic acid were also found. The study, conducted at the Biochemical Institute, University of Texas at Austin, concluded in summary: "We suggest that any real declines are generally most easily explained by changes in cultivated varieties between 1950 and 1999, in which there may be trade-offs between yield and nutrient content."

The debate surrounding genetic modification of plants is huge, encompassing the ecological impact of genetically modified plants, the safety of genetically modified food and concepts used for safety evaluation like Substantial equivalence.

Plant breeders' rights is also a major and controversial issue. Today, production of new varieties is dominated by commercial plant breeders, who seek to protect their work and collect royalties through national and international agreements based in intellectual property rights. The range of related issues is complex. In the simplest terms, critics of the increasingly restrictive regulations argue that, through a combination of technical and economic pressures, commercial breeders are reducing biodiversity and significantly constraining individuals (such as farmers) from developing and trading seed on a regional level. Efforts to strengthen breeders' rights, for example, by lengthening periods of variety protection, are ongoing.

Transgenic Plant

Transgenic plants are plants that possess a gene or genes that have been transferred from a different species. Such modification may be performed through ordinary hybridization through cross-pollination of plants, but the term today refers to plants produced in a labouratory using recombinant DNA technology in order to create plants with specific characteristics by artificial insertion of genes from other species, and sometimes entirely different kingdoms.

Prior to the current era of molecular genetics starting around 1975, transgenic plants, including cereal crops, were (since the mid 1930s) part of conventional plant breeding.

Transgenic varieties are frequently created by classical breeders who deliberately force hybridization between distinct plant species when carrying out interspecific or intergeneric *wide crosses* with the intention of developing disease resistant crop varieties. Classical plant breeder may use use of a number of *in vitro* techniques such as protoplast fusion, embryo rescue or mutagenisis to generate diversity and produce plants that would not exist in nature.

Such traditional techniques (used since about 1930 on) have never been controversial, or been given wide publicity except among professional biologists, and have allowed crop breeders to develop varieties of basic food crop, wheat in particular, which resist devastating plant diseases such as rusts. *Hope* is one such transgenic wheat variety bred by E. S. McFadden with a transgene from a wild grass. *Hope* saved American wheat growers from devastating stem rust outbreaks in the 1930s.

Methods used in traditional breeding that generate transgenic plants by non-recombinant methods are widely familiar to professional plant scientists, and serve important roles in securing a sustainable future for agriculture by protecting crops from pests and helping land and water to be used more efficiently.

Natural Movements of Genes

Natural movement of genes between species, often called horizontal gene transfer or lateral gene transfer, can occur because of gene transfer mediated by natural agent

This natural gene movement between species has been widely detected during genetic investigation of various natural Mobile genetic elements, such as transposons, and retrotransposons that naturally transfer to new locations in a genome, and often move to new species host over an evolutionary time scale. There are many types of natural mobile DNAs, and they have been detected abundantly in food crops such as rice.

These various mobile genes play a major role in dynamic changes to chromosomes during evolution, and have often been given whimsical names, such as Mariner, Hobo, Trans-Siberian Express (Transib), Osmar, Helitron, Sleeping Princess, MITE and MULE, to emphasize their mobile and transient behaviour.

Such genetically mobile DNA contitutute a major fraction of the DNA of many plants, and the natural dynamic changes to crop plant chromosomes caused by this natural transgenic

DNA mimics many of the features of plant genetic engineering currently pursued in the labouratory, such as using transposons as a genetic tool, and molecular cloning.

There is new scientific literature about natural transgenic events in plants, through movement of natural mobile DNAs called MULEs between rice and Setaria millet.

It is becoming clear that natural rearrangements of DNA and generation of transgenes play a pervasive role in natural evolution. Importantly many, if not most, flowering plants evolved by transgenesis - that is, the creation of natural interspecies hybrids in which chromosome sets from different plant species were added together. There is also the long and rich history of transgenic varieties in traditional breeding.

CREATION OF TRANSGENIC PLANTS

Production of transgenic plants in wide-crosses by plant breeders has been a vital aspect of conventional plant breeding for a century or so. Without it, security of our food supply against losses caused by crop pests such as rusts and mildews would be severely compromised. The first historically recorded interpecies transgenic cereal hybrid was actually between wheat and rye.

Last century, the introduction of alien germplasm into common foods was repeatedly achieved by traditional crop breeders by artificially overcoming fertility barriers. A novel genetic rearrangement of plant chromosomes, such as insertion of large blocks of rye (Secale) genes into wheat chromosomes ('translocations'), has also been exploited widely for many decades.

By the late 1930s with the introduction of colchicine, perennial grasses were being hybridized with wheat with the aim of transferring disease resistance and perenniality into annual crops, and large-scale practical use of hybrids was well established, leading on to development of Triticosecale and other new transgenic cereal crops. In 1985 Plant Genetic Systems (Ghent, Belgium), founded by Marc Van Montagu and Jeff Schell, was the first company to develop genetically engineered (tobacco) plants with insect tolerance by expressing genes encoding for insecticidal proteins from Bacillus thuringiensis (Bt).

Important transgenic pathogen and parasite resistance traits in current bread wheat varieties (gene, eg "Lr9" followed by the source species) are:

Disease resistance

- Leaf rust
  - Lr9 (from *Aegilops umbellulata*)
  - Lr18 *Triticum timopheevi*
  - Lr19 *Thinopyrum*
  - Lr23 *T. turgidum*
  - Lr24 *Ag. elongatum*
  - Lr25 *Secale cereale*
  - Lr29 *Ag. elongatum*
  - Lr32 *T. tauschii*

Fig. Leaf rust

- Stem rust
  - Sr2 *T. turgidum* ("Hope") McFadden, E. S., J. Am. Soc. Agron. 22, 1020-1031.
  - Sr22 *Triticum monococcum*
  - Sr36 *Triticum timopheevii*
- Stripe rust
  - Yr15 *Triticum dicoccoides*
- Powdery mildew
  - Pm12 *Aegilops speltoides*
  - Pm21 *Haynaldia villosa*
  - Pm25 *T. monococcum*
- Wheat streak mosaic virus
  - Wsm1 *Ag. elongatum*

Pest resistance

- Hessian fly
  - H21 *S. cereale* H23,

  - H24 *T. tauschii*
  - H27 *Aegilops ventricosa*
- Cereal cyst nematode
  - Cre3 (Ccn-D1) *T. tauschii*
- Lepidoptera
  - Bt *Bacillus thuringiensis*

The intentional creation of transgenic plants by laboratory based recombinant DNA methods is more recent (from the mid-80s on) and has been a controversial development opposed vigorously by many NGOs, and several governments, particularly within the European Community. These transgenic recombinant plants (= biotech crops, modern transgenics) are transforming agriculture in those regions that have allowed farmers to adopt them, and the area sown to these crops has continued to grow globally in each of the ten years since their first introduction in 1996.

Transgenic recombinant plants are now generally produced in a laboratory by adding one or more genes to a plant's genome, and the techniques frequently called transformation. Transformation is usually achieved using gold particle bombardment or a soil bacterium (*Agrobacterium tumefaciens*) carrying an engineered plasmid vector, or carrier of selected extra genes.

Transgenic recombinant plants are identified as a class of genetically modified organism (GMO); usually only transgenic plants created by direct DNA manipulation are given much attention in public discussions.

Transgenic plants have been deliberately developed for a variety of reasons: longer shelf life, disease resistance, herbicide resistance, pest resistance, non-biological stress resistances, such as to drought or nitrogen starvation, and nutritional improvement. The first modern transgenic crop approved for sale in the US, in 1994, was the FlavrSavr tomato, which was intended to have a longer shelf life. The first conventional transgenic cereal created by scientific breeders was actually a hybrid between wheat and rye in 1876. The first transgenic cereal may have been wheat, which itself is a natural transgenic plant derived from at least three different parenteral species.

Commercial factors, especially high regulatory and research costs, have so far restricted modern transgenic crop varieties to major traded commodity crops, but recently R&D projects to enhance crops that are locally important in developing counties are being pursued, such as insect protected cow-pea for Africa and insect protected Brinjal eggplant for India.

Regulation of Transgenic Plants

In the United States the Coordinated Framework for Regulation of Biotechnology governs the regulation of transgenic organisms, including plants. The three agencies involved are:

- USDA Animal and Plant Health Inspection Service

The Biotechnology Regulatory Services (BRS) programme of the U.S. Department of Agriculture's (USDA) Animal and Plant Health Inspection Service (APHIS) is responsible for regulating the introduction (importation, interstate movement, and field release) of genetically engineered (GE) organisms that may pose a plant pest risk.

BRS exercises this authority through APHIS regulations in Title 7, Code of Federal Regulations, Part 340 under the Plant Protection Act of 2000. APHIS protects agriculture and the environment by ensuring that biotechnology is developed and used in a safe manner.

Through a strong regulatory framework, BRS ensures the safe and confined introduction of new GE plants with significant safeguards to prevent the accidental release of any GE material. APHIS has regulated the biotechnology industry since 1987 and has authorized more than 10,000 field tests of GE organisms. In order to emphasize the importance of the programme, APHIS established BRS in August 2002 by combining units within the agency that dealt with the regulation of biotechnology.

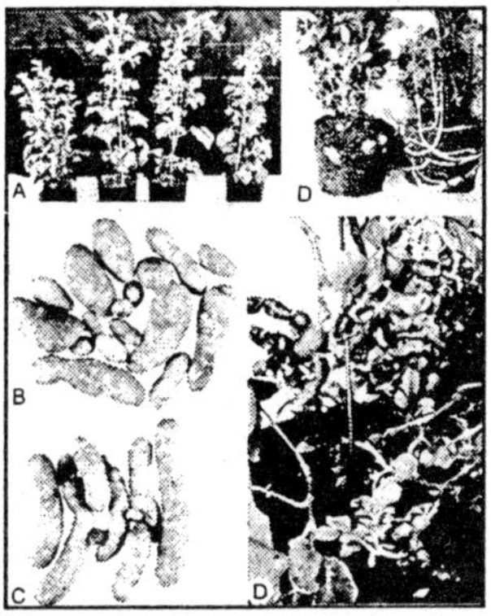

Fig. *Growth and Tuberization of Transgenic Plants with Altered Expression PCM-1 as Compared to Control. A: Control Plant on the left, and three independent Transgenic Plants to the right of control. Comparison of tuber shape and size B: (control), C: (transgenic). D: Tuberization pattern of one of the independent Transgenic Plants expressing the highest amount of PCM-1 (left, control; right, Transgenic plant). E: When the Transgenic Plant from figure D, was allowed to grow, aerial tubers were produced.*

Biotechnology, Federal Regulation, and the U.S. Department of Agriculture, February 2006, USDA-APHIS Fact Sheet

- EPA - evaluates potential environmental impacts, especially for genes which encode for pesticide production
- DHHS, Food and Drug Administration (FDA) - evaluates human health risk if the plant is intended for human consumption

Ecological risks

The potential impact on nearby ecosystems is one of the greatest concerns associated with transgenic plants.

Transgenes have the potential for significant ecological impact if the plants can increase in frequency and persist in natural populations. These concerns are similar to those surrounding conventionally bred plant breeds. Several risk factors should be considered:

- Is the transgenic plant capable of growing outside a cultivated area?
- Can the transgenic plant pass its genes to a local wild species, and are the offspring also fertile?
- Does the introduction of the transgene confer a selective advantage to the plant or to hybrids in the wild?

Many domesticated plants can mate and hybridise with wild relatives when they are grown in proximity, and whatever genes the cultivated plant had can then be passed to the hybrid. This applies equally to transgenic plants and conventionally bred plants, as in either case there are advantageous genes that may have negative consequences to an ecosystem upon release. This is normally not a significant concern, despite fears over 'mutant superweeds' overgrowing local wildlife: although hybrid plants are far from uncommon, in most cases these hybrids are not fertile due to polyploidy, and will not multiply or persist long after the original domestic plant is removed from the environment. However, this does not negate the possibility of a negative impact.

In some cases, the pollen from a domestic plant may travel many miles on the wind before fertilising another plant. This can make it difficult to assess the potential harm of crossbreeding; many of the relevant hybrids are far away from the test site. Among the solutions under study for this concern are systems designed to prevent transfer of transgenes, such as Terminator.

Technology, and the genetic transformation of the chloroplast only, so that only the seed of the transgenic plant would bear the transgene. With regard to the latter, there is some controversy that the technologies may be inequitable and might force dependence upon producers for valid seed in the case of poor farmers, whereas the latter has no such concern but has technical constraints that still need to be overcome.

There are at least three possible avenues of hybridization leading to escape of a transgene:

- Hybridization with non-transgenic crop plants of the same species and variety.
- Hybridization with wild plants of the same species.
- Hybridization with wild plants of closely related species, usually of the same genus.

However, there are a number of factors which must be present for hybrids to be created.

- The transgenic plants must be close enough to the wild species for the pollen to reach the wild plants.
- The wild and transgenic plants must flower at the same time.
- The wild and transgenic plants must be genetically compatible.

In order to persist, these hybrid offspring:

- Must be viable, and fertile.
- Must carry the transgene.

Studies suggest that a possible escape route for transgenic plants will be through hybridization with wild plants of related species.

1. It is known that some crop plants have been found to hybridize with wild counterparts.
2. It is understood, as a basic part of population genetics, that the spread of a transgene in a wild population will be directly related to the fitness effects of the gene in addition to the rate of influx of the gene to the population. Advantageous genes will spread rapidly, neutral genes will spread with genetic drift, and disadvantageous genes will only spread if there is a constant influx.
3. The ecological effects of transgenes are not known, but it is generally accepted that only genes which improve fitness in relation to abiotic factors would give hybrid plants sufficient advantages to become weedy or invasive. Abiotic factors are parts of the ecosystem which are not alive, such as climate, salt and mineral content, and temperature. Genes improving fitness in

relation to biotic factors could disturb the (sometimes fragile) balance of an ecosystem. For instance, a wild plant receiving a pest resistance gene from a transgenic plant might become resistant to one of its natural pests, say, a beetle. This could allow the plant to increase in frequency, while at the same time animals higher up in the food chain, which are at least partly dependent on that beetle as food source, might decrease in abundance. However, the exact consequences of a transgene with a selective advantage in the natural environment are almost impossible to predict reliably.

It is also important to refer to the demanding actions that government of developing countries had been building up among the last decades.

GENETICALLY MODIFIED FOOD CONTROVERSIES

The GM food controversy is a dispute over the advantages and disadvantages of genetically modified food crops. Although no major health hazards have come to light since GM food was introduced 13 years ago, and close to 150 studies are published to attest their safety. Consumer rights groups such as the Organic Consumers Association and Greenpeace emphasize the long term health risks which GM could pose, or that the risks of GM have not yet been adequately investigated.

Unnecessary delays to GM crop use by farmers pose another kind of risk. Agricultural scientist and economists express concern about the harm delaying welfare and environmental improvements, for instance by pro-vitamin A enriched Golden rice which has the potential to prevent much childhood death from infectious disease, and insect protected Bt rice which can reduce exposure of farmers to synthetic insecticides.

Safety disputes

Toxic GM-potatoes

In August 1998 widespread concern, especially in Europe, was sparked by remarks by nutrition researcher, Dr Arpad

Pusztai, regarding some of his research into the safety of GM foods.

Pusztai claimed his experiments showed that rats fed on potatoes genetically engineered to express a lectin from snowdrop had suffered serious damage to their immune systems and shown stunted growth. The lectin expressed by the genetically modified potatoes is toxic to insects and nematodes and is allegedly toxic to mammals. He was criticized by leading British politicians, the majority of scientific peers with expertise in the area and by the GM companies because the announcement of his results in a television interview, preceded the scientific publication of his results. When his studies were finally published in The Lancet, no evidence of stunted growth or damage to immune system was substantiated. The Lancet paper's actual summary was:

Fig. Toxic GM-potatoes

Diets containing genetically modified (GM) potatoes expressing the lectin *Galanthus nivalis* agglutinin (GNA) had variable effects on different parts of the rat gastrointestinal tract. Some effects, such as the proliferation of the gastric mucosa, were mainly due to the expression of the GNA transgene. However, other parts of the construct or the genetic transformation (or both) could also have contributed to the overall biological effects of the GNA-GM potatoes, particularly on the small intestine and caecum.

Fig. Galanthus nivalis

The paper was accompanied by an editorial explanation for allowing the paper's publication (Genetically modified foods: "absurd" concern or welcome dialogue?), and an independent critique which had a contrary evaluation of the data: Adequacy of methods for testing the safety of genetically modified foods. This was followed by a lively follow up debate in several later issues of the journal. Nonetheless, controversy about Pusztai's assertions still lingers, caused by strongly held but opposing views on his conclusions and data.

On the one hand, there are claims of misrepresentation of Pusztai's results by Rowett Research Institute, but on the other hand, there are concerns by scientists about overstatement of the quality of his findings by non-governmental organizations, and emphasis on matters well removed from the actual labouratory observations which are rarely discussed in public debate, against the context well over one hundred other studies published by 2006 that support the safety of GM foods and feeds, and commentaries such as that of Nina Fedoroff.

Research protocols were sent by Pusztai to 24 independent scientists in different countries (including experts in physiology, medicine, toxic pathology, nutrition, microbiology and biochemistry). These disagree with the conclusions of the review committee and argued that his research was of good quality and justified his conclusions. Among 'casualties' in these events was Dr Andrew Chesson, vice chairman of European Commission scientific committee on animal nutrition and former top scientist at the Rowett Institute who was fired for publicly defending Pusztai's research.

Various reports concerning the politicisation of the peer review process and alleged deliberate misrepresentation of Pusztai's results were voiced by newspapers and some scientists.

Dangerous Corn

Another controversy recently arose around biotech company Monsanto's data on a 90-Day Rat Feeding Study on a strain of GM corn. In May 2005, critics of GM foods pointed to differences in kidney size and blood composition found in this study, suggesting that the observed differences raises questions about the regulatory concept of substantial equivalence. Some argued that this study suggested human health might be affected by eating GM food.

However, the EU regulatory authorities that examined the Monsanto data concluded that the observed small numerical decrease in rat kidney weights was not biologically meaningful, and the weights were well within the normal range of kidney weights for control animals. There were no corresponding microscopic findings in the relevant organ systems, and all blood chemistry and organ weight values fell within the "normal range cf historical control values" for rats.

Thus, certain government authorities concluded that there were no effects on the functioning of kidneys in rats fed a diet of GM. Nevertheless, the debate about the Monsanto corn is not over. Genetics Professor Gilles-Eric Seralini, indicated in a scientific study that rats nourished with the ÌÏÍ863 Monsanto corn, showed toxicity signs in their livers and kidneys. The European Committee has approved the ÌÏÍ863 corn for animal and human consumption.

Allergenicity

A gene for an allergenic trait has been transferred unintentionally from the Brazil nut into genetically engineered soybeans while intending to improve soybean nutritional quality for animal feed use. Brazil nuts were already known to produce food allergies in certain people prior to this study. In 1993 Pioneer Hi-Bred International developed a soybean variety with an added gene from the Brazil nut.

This trait increased the levels in the GM soybean of the natural essential amino acid methionine, a protein building block commonly added to poultry feed to improve effective protein quality. Investigation of the GM soybeans revealed that they produced immunological reactions with people suffering from Brazil nut allergy, and the explanation for this is that the methionine rich protein chosen by Pioneer Hi-Bred is the major source of Brazil nut allergy. Pioneer Hi-Bred discontinued further development of the GM soybean and disposed of all material related to the modified soybeans.

While this study indicates the possible risks of GM foods, and indeed any new food source, some point out it establishes the commitment the developmental community has toward consumer safety as well as the competence of current safeguards. Food allergy problems occur with many conventional foods, and Kiwi fruit, for instance, as a relatively new food in many communities, has become widely eaten despite provoking allergies in certain individuals.

Another allergy issue was published in November 2005, when a pest resistant field pea developed by the Australian CSIRO for use as a pasture crop was shown to cause an allergic reaction in mice.

Respected plant scientist Maarten J Chrispeels has made interesting comments about this example that illustrate how foods offer many different types of risks:

The recent Prescott et al paper in JFAC contains a very interesting study on the immunogenicity of amylase [starch digestion enzyme] inhibitor in its native form (isolated from beans) and expressed as a transgene in peas. First of all, amylase inhibitor is a food protein, but also a "toxic" protein because it inhibits our digestive amylases. This is one of the reasons you have to cook your beans! (The other toxic bean protein is phytohemagglutinin and it is much more toxic).

This particular amylase inhibitor is found in the common bean (other species have other amylase inhibitors). Even though it is a food protein, it is unlikely ever to be used for genetic engineering of human foods because it inhibits our amylases. What the results show is that the protein, when synthesized in pea cotyledons has a different immunogenicity than when it is isolated from bean cotyledons (the native form). This is somewhat surprising but may be related to the presence of slightly different carbohydrate chains.

The immunologist who tested the pea noted that the episode illustrated the need for each new GM food to be very carefully evaluated for potential health effects.

Environmental and Ecological Impacts

As discussed above there is some evidence for positive impacts of the planting of GM crops on reduced greenhouse gas emissions and pesticide loads in the environment.

However, there has been controversy over the results of a farm-scale trial in the United Kingdom comparing the impact of GM crops and conventional crops on farmland biodiversity. Some claimed that the results showed that GM crops had a significant negative impact on wildlife.

Others pointed out that the studies showed that using herbicide resistant GM crops allowed better weed control and that under such conditions there were fewer weeds and fewer weed seeds. This result was then extrapolated to suggest that GM crops would have significant impact on the wildlife that might rely on farm weeds. In July 2005 the same British scientists showed that transfer of a herbicide-resistance gene from GM oilseed rape to a wild cousin, charlock, and wild turnips was possible.

Many agricultural scientists and food policy specialists view GM crops as an important element in sustainable food security and environmental management. This point of view is summarised in the ABIC Manifesto:

On our planet, 18% of the land mass is used for agricultural production. This fraction cannot be increased substantially. It is absolutely essential that the yield per unit of land increases beyond current levels given that: The human population is still growing, and will reach about nine billion by 2040;70,000 km^2 of agricultural land (equivalent to 60% of the German agricultural area) are lost annually to growth of cities and other non-agricultural uses; Consumer diets in developing countries are increasingly changing from plant-based proteins to animal protein, a trend that requires a greater amount of crop-based feed.

Public Perception

Research by the Pew Initiative on Food and Biotechnology has shown that in 2005 Americans' knowledge of genetically modified foods and animals continues to remain low, and their opinions reflect that they are particularly uncomfortable with animal cloning. The Pew survey also showed that despite continuing concerns about GM foods, American consumers do not support banning new uses of the technology, but rather seek an active role from regulators to ensure that new products are safe.

Interestingly, about 550 Amish farmers in Pennsylvania have adopted GM crops, because they allow for less intensive farming (less pesticides, etc.), are more productive (under these specific conditions), and do not conflict with the Amish lifestyle.

Opponents of genetically modified food often refer to it as "Frankenfood", after Mary Shelley's character Frankenstein and the monster he creates, in her novel of the same name. The term was coined in 1992 by Paul Lewis, an English professor at Boston College who used the word in a letter he wrote to the *New York Times* in response to the decision of the US Food and Drug Administration to allow companies to market genetically modified food. The term "Frankenfood" has become a battle cry of the European side in the US-EU agricultural trade war.

The authors of *The Frankenfood Myth* provide some support for genetically modified food:

- Henry I. Miller of Stanford's Hoover Institution and Gregory Conko of the Competitive Enterprise Institute make the case that foods modified by recombinant DNA splicing present no new or special dangers, but in fact may improve the lives of countless millions worldwide.

INTERNATIONAL TRADE OF GENETICALLY MODIFIED FOODS

The European Union and the United States have strong disagreements over the EU's regulation of genetically modified food. The US claims these regulations violate free trade agreements; the EU counter-position is that free trade is not truly free without informed consent.

In Europe, a series of unrelated food crises during the 1990s created consumer apprehension about food safety in general, eroded public trust in government oversight of the food industry, and left some consumers unwilling to consider "science" to be a guarantee of quality.

This has further fueled widespread public concern about genetically modified organisms (GMO), in terms of potential environmental protection (in particular biodiversity), health, and safety of consumers. Critics of GM foods contend that there is evidence that the cultivation of genetically modified plants may lead to environmental changes. Directives such as directive 2001/18/EC were designed to require authorisation for the placing GMO on the market, in accordance with the precautionary principle.

Many European consumers are demanding the right to make an informed choice about whether or not to consume GMO foods. Some polls indicate that Americans would also like labeling, but it has not yet become a major issue. New EU regulations should require strict labeling and traceability of all food and animal feed containing more than 0.5 percent GM ingredients.

A 2003 survey by the Pew Research Centre found that a majority of people in all countries surveyed felt that GM foods were "bad". The lowest scores were in the US and Canada, where 55% and 63% (respectively) were against it, while the highest were in Germany and France with 81% and 89% disapproving. The survey also showed a strong tendency for women to be more opposed to GM foods than men.

In 2002, Oregon Ballot measures gave voters in that state one of the first opportunities in the United States to directly address that issue. The measure, which would have required the labeling of genetically engineered foods, failed to pass by a ratio of 7 to 3.

Friedrich-Wilhelm Graefe zu Baringdorf, member of the German Green Party and vice president of the Landwirtschaftsausschuss (committee of agriculture) of the European Commission said on the 1 July 2003: "In America 55% of the consumers are against GM food and 90% in favour of a clear labeling."

The European Union and United States are in strong disagreement over the EU's ban on most genetically modified foods. The value of agricultural trade between the US and the European is estimated at $57 billion at the beginning of the 21st Century, and some in the U.S., especially farmers and food manufacturers, are concerned that the new proposal by the European Union could be a barrier to much of that trade.

In 1998, the United States exported $63 million worth of maize to the EU, but the exports decreased to $12.5 million in 2002.

The drop-off might also be due to falling commodities prices, less demand due to the recession, U.S. maize being priced out of foreign markets by a strong dollar, and importing countries' reaction to the planned invasion of Iraq. Similar European public opposition to Israeli treatment of Palestinians has also affected Israeli food exports. However, American farm industry advocates blame the EU's ban.

The European Parliament's Committee on the Environment, Public Health and Food Safety proposal, adopted in the summer of 2002 and expected to be implemented in 2003, has deep cultural roots that are difficult to understand for the US agricultural community. It requires that all food/feed containing or derived from genetically modified organisms be labeled and any GM ingredients in food be traced. It would also require documentation tracing biotechnological products through each step of the grain handling and food production processes.

The new European tax, tariff and trade proposal would particularly affect US maize gluten and soybean exports, as a high percentage of these crops are genetically modified in the USA (about 25 percent of US maize and 65 percent of soybeans are genetically modified in 2002).

The ultimate resolution of this case is widely thought to rest on labeling rather than food aid. Many European consumers are asking for food regulation (demanding labels that identify which food has been genetically modified), while the American agricultural industry is arguing for free trade and is strongly opposed to labeling, saying it gives the food a negative connotation.

Lori Wallach, director of Public Citizen's Global Watch indicates that American agricultural industry is "using trade agreements to determine domestic health, safety and environmental rules" because they fear that "by starting to distinguish which food is genetically modified, then they will have to distinguish energy standards, toxic standards that are different to those that European promotes."

The American Agricultural Department officials answer that since the United States does not require labeling, Europe should not require labeling either. They claim mandatory labelling could imply there is something wrong with genetically modified food, which would be also a trade barrier. Current U.S. laws do not require GM crops to be labeled or traced because U.S. regulators do not believe that GM crops pose any unique risks over conventional food. Europe answers that the labeling and traceability requirements are not only limited to GM food, but will apply to any agricultural goods. The American agricultural industry also complains about the costs implied by labeling.

The ban over agricultural biotechnology things is said by some Americans to breach World Trade Organisation rules. Robert B. Zoellick, the United States trade representative, indicated the European position toward GMO was thought of as "immoral" since it could lead to starvation in the [{developing country/developing world} of wars], as seen in some famine-threatened African countries (eg, Zambia, Zimbabwe, and Mozambique) that refuse to accept US aid because it contains GM food.

Zoellick's critics argue that US concern over Third World starvation is merely a cover for other issues. Some money for development aid is used by the American government via the World Food Programme (WFP) to help their farmers by buying up overproduction and giving it to the UN organisation. GM-scepticism interferes with this programme. American farmers lost marketshare in certain countries after changing to genetically modified food because of sceptical consumers.

Another European response to the claims of immorality is that the EU gives 7 times more in development aid than the US, yet its economy is less than 10% bigger than America's, and its GDP/head much lower than that of the US.

In May 2003, after initial delay due to the war against Iraq, the Bush administration officially accused the European Union of violating international trade agreements, in blocking imports of U.S. farm products through its long-standing ban on genetically modified food. Robert Zoellick announced the filing of a formal complaint with the WTO challenging the moratorium after months of negotiations trying to get it lifted voluntarily. The complaint was also filed by Argentina, Canada, Egypt, Australia, New Zealand, Mexico, Chile, Colombia, El Salvador, Honduras, Peru, and Uruguay. The formal WTO case challenging the EU's regulatory system was in particular lobbied by U.S. biotechnology giants like Monsanto and Aventis and big agricultural groups such as the National Corn Growers Association.

EU officials questioned the action, saying it will further damage trade relations already strained by the U.S. decision to launch a war against Iraq despite opposition from members of the U.N. Security Council. The US move was also interpreted as a sanction against the EU for requesting the end of illegal tax breaks for exporters or face up to $4 billion in trade sanctions in retaliation for Washington's failure to change the tax law, which the WTO ruled illegal four years ago.

Ratification of the Biosafety Protocol

In June 2003, the European Parliament ratified a three-year-old U.N. biosafety protocol regulating international trade in genetically modified food, expected to come into force in fall 2003 since the necessary number of ratification was reached in May 2003. The protocol lets countries ban imports of a genetically modified product if they feel there is not enough scientific evidence the product is safe and requires exporters to label shipments containing genetically altered commodities such as corn or cotton. It makes clear that products from new technologies must be based on the precautionary principle and allows developing nations to balance public health against economic benefits.

Jonas Sjoestedt, a Swedish Left member of the EU assembly, said that "this legislation should help the EU to counter recent accusations by the U.S administration that the EU is to blame for the African rejection of GM food aid last year." The United States did not sign the protocol, saying it was opposed to labeling and fought import bans.

In 1998, a *de facto* moratorium led to the suspension of approvals of new genetically modified organisms (GMO) in the European Union pending the adoption of revised rules to govern the approval, marketing and labelling of biotech products. Imports and cultivation of already approved GM varieties and food products continued. In July 2003, European environment ministers and the European Parliament agreed to new controls on GMOs that could eventually lead the then 15 members bloc to re-open the Union's markets to new genetically modified products in 2004.

The new labeling and traceability rules, which cover both food and feed, require any products with a GMO content of more than 0.9 percent to be labelled. Labelling is also required for products that have been derived from GMOs, but where the GM content might no longer be detectable (such as soy oil produced from genetically modified soy).

The threshold for the presence of unapproved GMOs is 0.5 percent provided that the GMOs have been judged as safe for human health and the environment by the relevant Scientific Committees or the European Food Authority. This amount will be set for 3 years. After 3 years, all food containing non-authorized GMO will be banned. Animals fed with transgenic cereals are not covered by the labeling requirements.

Traceability of GMO products is mandatory, from sowing to final product. Genetically modified goods will have to carry a special harmless DNA sequence (a DNA code bar) identifying the origin of the crops, making it easier for regulators to spot contaminated crops, feed, or food, and enabling products to be withdrawn from the food chain should problems arise. A series of additional sequences of DNA with encrypted information about the company or what was done to the product could also be added to provide more data.

Following the entry into force of the new regulations, the first genetically modified food product (canned maize) since 1998 was approved for marketing in the European Union in May 2004. While a number of other biotech products have been approved since then, approvals remain controversial. European ministers have continuously failed to reach a decision in support of or against the applications, highlighting the big divide among member states. As a result, the approvals were granted by the European Commission, which is entitled to take a decision in case ministers fail to do so.

The U.S. population has historically placed a considerable degree of trust in the regulatory oversight provided by the U.S. Department of Agriculture and its agencies. There is little tradition of people having a close relationship with their food, with the overwhelming majority of people having bought their food in supermarkets for years. But the 2003 survey by the Pew Research Centre showed that even in the U.S., 55% see GM food as "bad" food.

In Europe, and particularly in the UK, there is less trust of regulatory oversight of the food chain. In many parts of Europe, a larger measure of food is produced by small, local growers using traditional (non-intensive & organic) methods.

Arpad Pusztai, considered by many to be the leading expert on GM foods, was silenced with threats of a lawsuit after he unexpectedly discovered that rats fed an experimental GM food developed immune system damage and other serious health problems in just ten days. Pusztai later reviewed an industry-sponsored study and found that seven of forty rats fed a GM crop died within two weeks; others developed stomach lesions. The crop was approved without further tests.

In May, when the U.S. filed a challenge with the World Trade Organization (WTO) disputing Europe's GM food policy, Trade Representative Robert Zoellick stated, "Overwhelming scientific research shows that biotech foods are safe and healthy." According to Andrew Kimbrell, director of the Centre for Food Safety, "The evidence in the book Seeds of Deception refutes U.S. science and safety claims, and undermines the basis of their WTO challenge."

Kimbrell says, "Author Jeffrey M. Smith's book also presents a compelling argument that nations may use to ban GM foods altogether." Countries gained the right to impose such a ban on September 11, three months after the UN biosafety protocol was signed by 50 nations. "The revelations in the book," says Kimbrell, "are being made public at a pivotal time in the global GM debate, and could tip the scales against the biotech industry."

The World Trade Organization has made a preliminary ruling that European Union restrictions on genetically engineered crops violate international trade rules. The United States, Canada, and Argentina together grow 80 percent of all biotech crops sold commercially, by which the EU regulates such crops. The countries argued that the EU's regulatory process was far too slow and its standards were unreasonable given that the overwhelming body of scientific evidence finds the crops safe.

10

Future of Agriculture Biotechnology

COMES OF AGE

Transgenic crop plants — those with a genetic trait introduced by a molecular technique — have begun to make the journey from laboratory to field. Tomatoes with improved quality traits such as longer shelf-life, oil-seed rapes that are more tolerant of herbicides, and virus-resistant potatoes are only three examples of the many innovations that will soon appear in commercial farming. The likely extent of commercialisation will be vast; the economic consequences will also be enormous. But can one be sure that these plants will be as safe in an open environment as their traditional counterparts? Over the last several years, the advance of plant biotechnology towards commercialisation has speeded up at a remarkable rate.

More than 1,180 small-scale field trials with transgenic plants were conducted in the OECD area between 1986 and 1992; the number of trials has nearly doubled each year. The aim of researchers in these trials is to assess, in an open environment, the performance (virus resistance, for example) and environmental behaviour (say, the distance to which pollen spreads) of transgenic plants which have been developed in laboratories and tested in greenhouses. The trials involved around 30 different kinds of crops.

Oilseed rape, potato, maize, soybean, tobacco, cotton and tomato are the: most commonly used 'hosts' (crop plants into which a gene is inserted), and account for more than 80% of the total trials. The traits most commonly tested are resistance to herbicides, viruses and insects, and quality traits. Herbicide resistance alone accounts for 40% of the total number of traits tested. The trials were conducted mainly in the United States, Canada, the United Kingdom, France, India and Belgium, which accounted for 95% of the total in the area.

Because of differing, national circumstances, biosafety issues can vary, for valid scientific reasons. In the United States, for example, the major crop plants — maize, soybean, wheat and cotton — have virtually no wild relatives, which would make the issue of 'outcrossing' (uncontrolled interbreeding between domesticated and wild species) unimportant for these crops. For Norway, by contrast, with strong economic dependence on native trees, it would be important to ensure that introduction of traits into related species of tree does not adversely influence the long-term reliability of this natural resource.

In commercial production, which is imminent and will be on a huge scale, crop plants will be grown in a completely open environment. But small-scale trials are limited in their ability to predict certain kinds of events that may arise in large-scale cultivation in the open. For example, a small-scale trial may not Seizo Sumida works in the "Biotechnology Unit of the Science and Technology Policy Division of the OECO Directorate for Science".

BIOTECHNOLOGY SAFETY

It was the introduction of new molecular techniques in the early 1970s that initiated the discussion on safety in biotechnology. The 1986 OECD report on the safety considerations of recombinant DNA was one of the first international scientific frameworks for the safe use of organisms derived from the technique in industry, agriculture and the environment, and the OECD countries adopted the scientific principles into their regulatory systems.

In the light of rapid accumulation of experience and knowledge in the field, the OECD resumed its safety work in

April 1988 to update and extend the particular principles set out in the earlier report. The result was a set of developmental principles for small-scale field trials of genetically modified organisms, published in 1992. The OECD further initiated, in 1991, similar activities on large-scale field trials, completing its work on transgenic plants in early 1993.

The work produced three reports: safety considerations for large-scale trials of crop plants, analysis of small-scale field trials and a historical review of crop breeding practices, published this autumn.

Large-scale trials, conducted, for example, over a wider geographical area, may be the only means of obtaining the data to answer the questions of plant safety. In order to tackle these issues, the OECD has developed a new set of scientific principles for the environmental safety of the 'scale-up' of crop plants. These principles are intended as guidelines primarily for the regulatory agencies, scientists and industries responsible for the authorisation of the environmental release of crop plants in OECD countries. (The system for authorisation varies considerably from country to country.)

Under these guidelines, a transgenic plant under consideration is first examined by the relevant authorities for 'safety issues'. A 'safety issue' is some property of a plant (say, the possibility of gene transfer to a wild relative by out crossing, the possibility that the plant becomes a weed, and so on) which may or may not give rise to an adverse effect in the environment. When such a factor is directly identified in a given environment, it gives rise to a 'safety concern'.

For example, out crossing between herbicide-resistant transgenic sunflower and its wild relative could be a 'safety concern' in some areas of the United States. The identification of such a concern indicates where the analysis should then focus, and what form risk management should take. Standard agronomic practices in crop cultivation may form part of risk-management, because crops are so domesticated that they usually cannot compete with wild plants outside the plot or field. Maize, for example, cannot survive without human help (that is, it cannot 'escape' from the cultivated field).

The knowledge and experience gained with crop plants developed by traditional breeding methods are essential to address the issues of environmental safety. This kind of experience can guide risk/safety analysis, applying risk-management and indicating where more information is required, and the more that is known about a given plant, its traits, its environment and their likely interactions, the easier risk/safety-analysis and risk-management become.

In the small-scale field trials, there have so far been no surprises in the behaviour of the transgenic plants, in relation to what might be predicted from the crop host and the genetic trait. At the moment large-scale field trials have yet to be conducted. The OECD will be providing guidance to the agencies and others involved in authorising or carrying out these large-scale trials.

There is a substantial difference in the views on safety assurance between the European Community countries and the US and Japanese governments. Directives issued by the EC authorities have insisted that all transgenic organisms be regulated. The US and Japanese governments emphasise that the issue, rather, is whether the final products offer a risk, regardless of how the organism was modified. This difference in approach was evident when the OECD work on large-scale releases of transgenic crop plants started in 1991. But, after a couple of years of intense discussion, all the OECD countries have reached consensus at least on the scientific principles underlying the safety of the large-scale releases.

The transgenic organisms are now held to present risks that are of the same basic sort as those posed by any other organism. This agreement reflects changing scientific, technological and economic circumstances. OECD governments have recognised that experience of transgenic plants has been accumulating rapidly, that evolution of modern biotechnology is dynamic and fast, and that they have to adapt their positions according to new circumstances if they are to maintain the basis for further development of their agricultural and food sectors. This trend will also facilitate international R&D co-operation, investment and trade.

The possibilities of plant biotechnology are enormous. The regeneration of a whole plant from a piece of cellular tissue is now common. The insertion of a new gene into a plant by molecular techniques is possible in many species. What is more, technical evolution is extraordinarily swift, which means that safety issues will have to be revisited, and scientific principles updated, as knowledge and experience in large-scale field releases accumulate.

A Nitrogen Fix

Some plants can find their own fertilizer, but corn isn't one of them. It can cost farmers $40 per acre to fertilize a cornfield. Wisconsin scientists went looking for bacteria that live in corn and can capture nitrogen from the air. Greenhouse experiments narrowed the field to seven strains, and field tests showed the bacteria improved corn yields up to 10 percent. Several companies are interested in licensing the technology, which could result in seed corn coated with growth-promoting bacteria and reduced nitrogen runoff into streams and rivers.

Florida researchers found a gene in pond algae that helps the microscopic plants compete better for nitrogen. They then worked with a biotech company to create a transgenic wheat plant that produced significantly more grain than conventional wheat for the same amount of fertilizer.

Booster Shots

Despite some natural resistance, barley is a pushover for stem rust. Minnesota scientists genetically enhanced that resistance, the first time a gene for rust resistance has been isolated from a small-grain cereal crop. Not only is this genetically engineered resistance better than the original barley plant, scientists think it also may work in wheat. Fire blight, a bacterial disease in apple trees, annually costs growers more than $100 million. It can be treated only with the antibiotic streptomycin, but not for long - some bacteria already are resistant. Cornell researchers are using biotechnology to enhance the tree's existing genes to help it fight fire blight, greatly reducing tree death.

Health Benefits

Breast cancer-fighting taxol once was available only from slow-growing Pacific yew trees, and efforts to manufacture the compound have been problem-ridden. Washington State scientists have tracked the biological pathway the tree follows to produce the drug and have identified about half the genes involved. Once the work is complete they hope to be able to produce more taxol at lower cost and investigate new taxol derivatives with greater potency and fewer side effects.

Georgia researchers used enzyme biotechnology to attach beneficial fatty acids to conventional fats and oils, the least recommended part of the food pyramid. Labouratory trials with mice showed the designer fats lowered cholesterol and bolstered immune system cells, an encouraging outcome for people with AIDS and for other immuno-suppressed people.

Hot Chocolate

Chocolate manufacturers in Pennsylvania have the milk they need to support their $5 billion retail industry, but imported cocoa is hard to come by, with 40 percent of the crop lost to insects and disease each year. Pesticides help but they're expensive and nobody wants them near the rain forests where most cocoa is grown. Traditional plant breeding for resistance is slow and uncertain. Starting with plants endowed with superior flowers and pest resistance, Penn State researchers cloned individual cells and grew them into full-size cocoa plants.

It's now possible to clone as many as 4,000 plants from just one flower, and the university has worked with scientists in seven cocoa-producing countries to make sure they're comfortable with the technology.

The $45 million Texas citrus industry is at risk from two highly infectious plant diseases that have caused widespread damage around the world. Texas A&M scientists have moved genes into Red Rio grapefruit that helps protect the trees against citrus tristeza virus, brown aphids and citrus canker, a bacterial disease. The genetic protection comes from a vaccine derived from the original virus, an insect destroying protein from a lily commonly found in the northeastern United States and a milk gene that battles bacteria.

Advancing animal well-being may get some help from research that identifies the genetic basis for aggressive behaviour. Purdue researchers showed that breeding swine and poultry to cooperate rather than compete can improve productivity and decrease mortality. An experimental line of quail bred to behave had 25 percent better feed efficiency. In pigs, 20 percent more growth for the same amount of feed could increase net incomes by $2 billion.

Veterinary medicine and animal disease diagnosis have improved thanks to new genetic technologies that speed vaccine and diagnostic tool development. Tennessee researchers devised several antibodies that detect a substance called antigen 85 in cows infected with Johne's disease, one of the top three diseases in beef and dairy cattle and responsible for $250 million annual economic losses.

They hope Ag85 helps them develop a vaccine and a better diagnostic kit. Scientists at the Virginia-Maryland Regional College of Veterinary Medicine developed a livestock vaccine against brucellosis, a disease that affects both animals and humans. The U.S. Army has asked researchers to develop a human vaccine as well. These tools would become even more critical in the event of a bioterrorism incident.

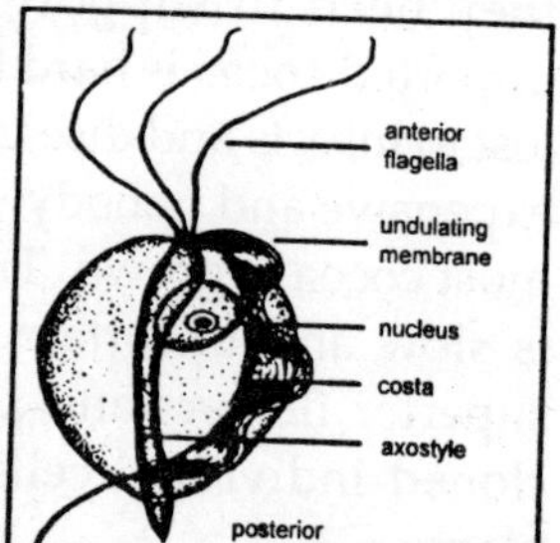

Fig. Tritrichomonas foetus

Basic genetic research helped Nevada animal scientists develop a vaccine for *Tritrichomonas foetus*, a parasite that causes a reproductive disease in cattle. The vaccine is the only pretreatment available for the disease and may save Nevada cattle ranchers \$950,000 per year. Western U.S. cattle producers also are looking forward to another Nevada project that will detect epizootic bovine abortion, a tick-borne disease that causes cows to lose their unborn calves at six to nine months. The test is quicker, better and cheaper and may lead to new treatments. Officials estimate this disease may kill up to 90,000 calves annually in California alone.

The late sixties and seventies witnessed Indian plant breeders taking giant strides in enhancing wheat and rice productivity, transforming India from an importing nation to one exporting food grains. More recently rate of gain in productivity seems to have plateaued. There were times when breeders looked at the emerging fields of mutation breeding and tissue culture with awe and expectations. The illusion is over and we are witness to the real picture. These have occupied their place in the history of development of science and/or as an adjunct to the major field, the discipline nuclear to crop science research, i.e. plant breeding.

Indian plant biotechnology has come of age accomplishing research projects of national and international importance, e.g. rice genome sequencing project. Plant Biotechnology (PB) offers two major options to plant breeders. Marker- assisted selection (MAS) offers to make selection for desirable segregants precise and expression independent 1. The question, however, which traits and who will do it, remains unattended. The molecular biologists who have so far been experimenting with it are alienated from those who will ultimately be practising MAS. The moment science of MAS for a trait of importance is perfect enough to be- come a technology, the same needs to be transferred to the end users, the plant breeders in this case.

The interesting aspect of plant biotechnology outputs is that they need to pass through plant breeders before they reach the final consumers. Molecular biologists who tag a trait, need to be encouraged to convert it into a technology for use by breeders. Research managers can play a role to en- sure that funds invested in these scientific endeavours lead to usable technology and the same is passed on to breeders to cut down the enormous costs involved in the elabourate plant breeding operations. The next logical question is to decide which traits to tag? The obvious answer is those which breeders find difficult to select for. Transgenic technology is the other major important intervention that PB offers.

This tool has immense potential and the same is evident from the fact that currently over 130 million acres are planted under transgenic crops the world over. The global market value of biotech crops, which stood at 3.8 billion US dollars in 1998, is slated to rise to five billion US dollars by 2005.

India has also benefited from this technology by adopting boll worm-resistant transgenic cotton. This is one field, which needs to be strengthened by investing human and financial resources in the form of groups dedicated to specific trait and crop. Development of technology needs to be regarded as an equally important contribution as publication. Only such an approach will encourage researchers to be focused and dedicated to the product development rather than just the publication, which in turn will make PB more relevant and responsive to the society's needs.

Biotechnology has transformed many parts of the chemical industry, agriculture, and medicine. This area of science has little demarcation between basic and applied research, and new discoveries and innovations, in most cases, can find direct application. Innovations, techniques, and tools that have emerged and revolutionized modern biotechnology include genetic engineering, cell fusion technology, bioprocess technologies, and structure-based molecular designs including drug development, drug targeting, and drug delivery systems.

In the 1980s the Government of India considered the need for creating a separate institutional framework to strengthen biology and biotechnology research in the country. Scientific agencies supporting research in modern biology included: Council of Scientific and Industrial Research (CSIR), Indian Council of Agricultural Research (ICAR), Indian Council of Medical Research (ICMR), Department of Science and

Technology, and University Grants Commission. Biotechnology was given an important boost in 1982 with the establishment of the National Biotechnology Board. Its priorities were human resource development, creation of infrastructure facilities, and supporting research and development (R&D) in specific areas.

The success and impact of the National Biotechnology Board prompted the Government to establish a separate Department of Biotechnology (DBT) in February 1986. There have been major accomplishments in areas of basic research in agriculture, health, environment, human resource development, industry, safety, and ethical issues.

Basic Research

Basic research is essential on all aspects of mod- ern biology including development of the tools to identify, isolate, and manipulate the individual genes that govern the specific characters in plants, animals, and microorganisms. Recombinant DNA (rDNA) technology is the basis for these new developments. The creativity of the scientists and the basic curiosity-driven research will be the keys to future success. India led through the work of G.N. Ramachandran, in which he elucidated the triple helical structure of collagen. The Ramachandran plot has proven to be fundamental in solving the protein structure. Areas of biosystematics using molecular approaches, mathematical modeling, and genetics including genome sequencing for human beings, animals, and plants, will continue to have priority as we move into the next century.

The tremendous impact of genome sequencing is increasingly evident in many fields. As an increasing number of new genes are discovered, short, unique, expressed sequenced tags segments are used as signatures for gene identification. The power of high throughput sequencing, together with rapidly accumulating sequenced data, is opening new avenues in biosciences.

In the plant genome area, the sequencing of *Arabidopsis* and rice genome will soon be completed and cataloging and mapping of all the genes will be done. There have been major achievements in basic bioscience in the last decade or so in India, where we have expertise in practically all areas of modern biology. The institutions under the CSIR, ICMR, ICAR, DST, and DBT have established a large number of facilities where most advanced research work in biosciences is being done. In the identification of new genes, development of new drug delivery systems, diagnostics, recombinant vaccines, computational biology, and many other related areas, considerable success has been achieved.

Breakthroughs include studies on the three-dimensional structure of a novel amino acid, a long protein of mosquito (University of Poona), and demonstration of the potential of the reconstituted Sendai viral envelops containing only the F protein of the virus, as an efficient and site-specific vehicle for the delivery of reporter genes into hepatocytes (Delhi University).

Agriculture and Allied Areas

The post Green Revolution era is almost merging with the gene revolution for improving crop productivity and quality. The exploitation of heterosis vigour and development of new hybrids including apomixis, genes for abiotic and biotic resistance, and developing planting material with desirable traits and genetic enhancement of all important crops will dominate the research agenda in the next century. Integrated nutrient management and development of new biofertilizers and biopesticides would be important from the view- point of sustainable agriculture, soil fertility, and a clean environment.

Stress biology, marker-assisted breeding programmes, and studying the important genes will continue as priori- ties. We will have to switch to organic farming practices, with greater use of biological software on a large scale.

In India we have achieved the cloning and sequencing of at least six genes, developed regeneration protocols for citrus, coffee, mangrove species, and new types of biofertilizer and biopesticide formulations, including mycorrhizal fertilizers. Research to develop new genetically improved (transgenic) plants for brassicas, mung bean, cotton, and potato is well advanced. Indus- tries have also shown a keen interest in the options of biotechnology and are participating in field trials and pilot level productions.

The successful tissue culture pilot plants in the country, one at TERI in New Delhi and the other at NCL in Pune are now functioning as Micropropagation Technology Parks. This has given a new direction to the plant tissue culture industry. The micropropagation parks serve as a platform for effective transfer of technology to entrepreneurs, including training and the demonstration of technology for mass multiplication of horticulture and trees. Considerable progress has been made with cardamom and vanilla, both important crops.

Yield of cardamom has increased 40 percent using tissue-cultured plants. Between 1996 and 1998, in just eight countries, the area covered by new genetically improved transgenic plants (from 16.8 to 27.8 million hectares). Some of the main crops grown are soybean, corn, canola, cotton, and potato. The United States, Argentina, Brazil, and China have moved ahead quickly. The new plants exhibited herbicide, insect, and viral resistance, and over- all improvement in product quality. While the Green Revolution gave us self-reliance in food, the livestock population has pro- vided a "White Revolution," with 80 percent of the milk in India coming from small and marginal farms. This has had a major social impact.

A diverse infrastructure has been established to help farmers in the application of embryo transfer technology. The world's first IVF buffalo calf (PRATHAM) was born through embryo transfer technology at the National Dairy Research Institute, Karnal. Multiple ovulation and embryo transfer, in vitro embryo production, embryo sexing, vaccines and diagnostic kits for animal health have also been developed. Waste recycling technologies that are cost effective and environmentally safe, are being generated. The animal science area is also opening up many avenues for employment generation.

With a coastline of more than 8,000 kilometers, and two island territories of Andaman and Nicobar and Lakshadweep, there is great potential for marine resource development and aqua- culture. To achieve an annual target production of 10 million metric tons of fish, scientific aquaculture offers great possibilities. In fact, aquaculture products are among the fastest moving commodities in the world. We have to continuously improve seed production, feed, health products, cryopreservation, genetic studies, and related environmental factors. This is an area which will help substantially in the diversification of the breadbasket, and in combating nutritional deficiency.

Food Security

Food security is another area in which biotechnology offers major inputs for healthier and more nutritious food. Millions of people are malnourished, and Vitamin A deficiency affects 40 mil- lion children. There are also serious deficiencies of iodine, iron, and other nutrients. A recent UNICEF report on food and nutrition deficiencies in children describes this as a "silent, invisible emergency with no outward sign of a problem." Every year over 6 million children under the age of 5 die worldwide. About 2.7 mil- lion of these children die in India. More than half of these deaths result from inadequate nutrition. With the advent of gene transfer technology and its use in crops, we hope to achieve higher productivity and better quality, including improved nutrition and storage properties.

We also hope to ensure adaptation of plants to specific environmental conditions, to increase plant tolerance to stress conditions, to increase pest and disease resistance, and to achieve higher prices in the marketplace. Genetically improved foods will have to be developed under adequate regulatory processes, with full public understanding. We should ensure the safety and proper labeling of the genetically improved foods, so consumers will have a choice.

It is scientifically well established that an environmentally benign way of ensuring food security is through bioengineering of crops. For the 4.6 billion people in developing countries, one billion do not get enough to eat and live in poverty. Is there any other strategy or alternative? Biotechnology will provide the new tools to breeders to enhance plant capacity. Since we know that 12 percent of the world land is under agricultural crops, it is projected that the per capita availability may be reduced from 2.06 hectares to 0.15 hectare by 2050.

Plant Biotechnology

With more than 47,000 species of plants and two hot-spots of biodiversity, 8 percent of the total biodiversity of the earth is available in the Indian subcontinent. The bioresource and biodiversity constitute the mainstay of the economy of the poor people, and special emphasis is required for plant biotechnology research. Isolation of genes for abundant proteins, combining molecular genetics and chromosome maps, and a much better understanding of the evolutionary relationship of the members of the plant kingdom, have led to the potential of plant species being the major source of food, feed, fiber, medicine, and industrial raw material.

Molecular fingerprinting and areas of genomics and proteonics will penetrate the barriers of fertilization to allow transfer of important characters from one plant to another. By identifying appropriate determinants of male sterility, we can extend the benefit of hybrid seeds to more crops. We must help the farmer by ensuring hybrid vigour generation after generation. Additional research on apomixis would open up such possibilities.

We have set up a National Plant Genome Re- search Centre at Jawaharlal Nehru University. A number of centers for plant molecular biology in different parts of the country were initially responsible for training significant numbers in crop biotechnology. There are innumerable possibilities of producing more proteins, vitamins, pharmaceuticals, coloring material, bioreactors, production of edible vaccines, therapeutic anti- bodies and drugs. Promising leads are available in these areas, and a number of genetically improved crops are ready for field trials of transgenic plants. Work on developing transgenic cotton, brassica, mung bean, and potato has significantly advanced.

Environment

A special area of global concern amongst the scientific community is environmental protection and conservation, and the need for a policy of sustainable development in harmony with the environment. The Stockholm Conference in 1972, and the UNCED Conference in Rio de Janeiro in 1992, both focused world attention on areas of pollution, biodiversity conservation, and sustain- able development. Plants and microbes are becoming important factors in pollution control. World Bank estimates show that pollution in India is costing almost US$80 billion, as well as the human cost in terms of sickness and death.

New developments such as bioindicators, phytoremediation methods, bioleaching, development of biosensors, and identification and isolation of microbial consortia are priority research areas. Significant work has been done in India, but developing a more biologically oriented approach towards pollution control would be extremely important. Cleaning up the large river systems and ensuring the destruction of pesticide residue in large slums in the city are priorities in which a biotechnological approach would be environmentally safe.

Phytoremediation to remove the high levels of explosives found in the soil has become a reality. Although it was known that some microbes can denitrify the nitrate explosives in the laboratory, they could not thrive on site. French and others have transferred this degradative ability from the microbe to tobacco plants, and these have produced a microbial enzyme capable of removing the nitrates.

Biodiversity

The global biosphere can survive only if resource utilization is about 1 percent and not 10 percent. The global environment is regulated by climate changes and biosphere dynamics. Knowledge about biodiversity accumulated in the last 250 years is being used by scientists throughout the world. There are many gene banks, botanical gar- dens, and herbaria for conservation purposes. There are also molecular approaches including DNA fingerprinting for plant conservation. The totality of gene species and ecosystems has become exceedingly important, not only to under- stand the global environment but also from the viewpoint of the enormous commercial significance of the biodiversity.

Biotechnology is becoming a major tool in conservation biology. Twelve percent of the vascular plants are threatened with extinction. Over 5,000 animal species are threatened worldwide, including 563 Indian species. India also has about 2000 species of vascular plants that are threatened. Biodiversity is under threat, and understanding the scale of this destruction and extinction is essential. Questions such as who owns the biodiversity, who should benefit from it, and what is the role of society and the individual are pertinent. There is a Kashmiri proverb that says: *We have not inherited the world from our forefathers, we have borrowed it from our children.*

More research is needed on forests, marine re- sources, bioremediation methods, restoration ecology, and large-scale tree plantations. The last has reached 180 million hectares and may increase substantially in the next decade. Marine resources provide many goods and benefits including bioactive materials, drugs, and food items and must be characterized and conserved.

Medical Biotechnology

A major responsibility of biotechnologists in the 21st century will be to develop low-cost, afford- able, efficient, and easily accessed health care systems. Advances in molecular biology, immunology, reproductive medicine, genetics, and genetic engineering have revolutionized our understanding of health and diseases and may lead to an era of predictive medicine. Genetic engineering promises to treat a number of mono- genetic disorders, and unravel the mystery of polygenetic disorders, with the help of research on genetically improved animals.

Globally, there are about 35–40 biotechnology-derived therapeutics and vaccines in use and more than 500 drugs and vaccines in different stages of clinical trials. Every year about 12 million people die of infectious diseases. The main killers according to WHO are acute respiratory infection, diarrhoea diseases, tuberculosis, malaria, hepatitis, and HIV-AIDS. There are vaccines being developed for many diseases, and diagnostic kits for HIV, pregnancy detection, and hepatitis are being developed. The technologies have been transferred to industry.

The Department of Biotechnology has developed guidelines for clinical trials for recombinant products, which have now been accepted by the Health Ministry and circulated widely to industry. Promising leads now exist to develop vaccines for rabies, *Mycobacterium* tuberculosis, cholera, JEV, and other diseases. Recombinant hepatitis B vaccine and LEPROVAC are already on the market.

There is a Jai Vigyan technology mission on the development of vaccines and diagnostics. A National Brain Research Centre is being established to improve knowledge of the human brain and the brain diseases. The discovery of new drugs and the development of the drug delivery system are increasingly important. Bioprospecting for important molecules and genes for new drugs has begun as a multi-institutional effort. A recombinant vaccine for BCG and hepatitis is being developed. The age-old system of Ayurveda practiced in India needs to be popularized and made an integral part of health care. The global market for herbal products may be around US$5 trillion by 2050.

Industrial Biotechnology

Advances in biotechnology can be converted into products, processes, and technologies by creating an interdisciplinary team. The pharmaceutical sector has had a major impact in this field, as rare therapeutic molecules in the pure form be- come available. Diagnostics have expanded, with over 600 biotechnology-based diagnostics (valued at about US$20 billion worldwide) now avail- able in clinical practice. The polymerase chain reaction (PCR)-based diagnostics are the most common.

Indian efforts in the diagnostic area have been commendable, and it is expected that sales will rise from about US$235 million to US$470 million in the next century. The consumption of biotechnology products is expected to increase from US$6.4 billion to about US$13 billion by 2000. Industrial enzymes have emerged as a major vehicle for improving product quality.

In India a number of groups are gearing up to produce industrial enzymes such as alpha-amylose, proteases, and lipases, increasing three-fold by the end of the century, which will match or surpass the computer industry in size, importance, and growth. India is now producing 13 antibiotics by fermentation. Capacity exists to produce important vaccines such as DPT, BCG, JEV, cholera, and typhoid. Cell culture vaccines such as MMR and rabies, and hepatitis-B, have also been introduced

Bioinformatics

The coming together of biotechnology and informatics is paying rich dividends. Genome projects, drug design, and molecular taxonomy are all becoming increasingly dependent on in- formation technology. Information on nucleotides and protein sequences is accumulating rapidly. The number of genes characterized from a variety of organisms and the number of evolved protein structures are doubling every two years.

DBT has established a national Bioinformatics Net- work with ten Distributed Information Centres (DICs) and 35 sub-DICs. AJai Vigyan Mission on establishment of genomic databases has been started, with a number of graphic facilities created throughout the country. This system has helped scientists involved in biotechnology re- search.

Ethical and Biosafety Issues

The bioethics committee of UNESCO established in 1993 has evolved guidelines for ethical issues associated with the use of modern biotechnology. Biosafety guidelines for genetically improved organisms (GIOs) need to be strictly followed to prevent harm to human health or the environment. A three-tier mechanism of Institutional Biosafety Committees has been instituted in India: the Review Committee on Genetic Manipulation, the Genetic Engineering Approval Committee, and the state level coordination committees.

It is important to give a clear explanation of the new biotechnologies to the public to allay their fears. New models of cooperation and partnership have to be established to ensure close linkages among research scientists, extension workers, industry, the farming community, and consumers.

Gene transformation is done worldwide with four broad objectives:

1. To develop products with new characteristics;
2. To develop pest and disease resistance;
3. To improve nutritional value;
4. To modify fruit ripening to obtain longer shelf life. Thus the aims and objectives are laudable and the tools are available. The new technology does, however, call for a cautious approach following appropriate biosafety guide- lines. About 25,000 field trials of genetically modifed crops have been conducted worldwide. The anticipated benefits are better planting material, savings on inputs, and genes of different varieties can be introduced in the gene pool of crop species for their improvement. The potential risks include weediness, transgene flow to nontarget plants, and the possibility of new viruses developing with wider host range and their effects on unprotected species. For crops such as corn and cotton with single gene introductions, there is very little problem expected. When multiple genes are involved scientists have to be more cautious. The time has arrived for a serious look at ethical and biosafety aspects of biotechnology. Re- searchers, policymakers, NGOs, progressive farmers, industrialists, government

representatives, and all concerned players need to come together and share a platform to address the following issues.

- Environmental safety
- Food and nutrition security
- Social and economic benefits
- Ethical and moral issues
- Regulatory issues.

Human Resource Development

There are about 50 approved MS, postdoctoral, and MD training programmes in biotechnology in progress or just about to start, in different institutions and universities covering most Indian States. Short-term training programmes, technician training courses, fellowships for students to go abroad, training courses in Indian institutions, popular lecture series, awards, and incentives form an integral part of the human resource development activities in India.

A special feature of the programme has been that since 1996 many students after completion of their training course join industries or work in biotechnology-based pro- grams in institutions and laboratories. National Bioscience Career Development Awards have been instituted. Special awards for women scientists and scholarships to the best students in biology help promote biotechnology in India and give recognition and reward to the scientists.

Some Special Programmes

Biotechnology-based activities to benefit the poor and weaker sections and programmes for women have been launched. A unique feature is the establishment of a Biotechnology Golden Jubilee Park for Women which will encourage a number of women entrepreneurs to take up biotechnology enterprises that benefit women in particular.

This will also encourage women biotechnologists to develop relevant technologies. States are taking a keen interest in developing biotechnology-based activities. The States of

Uttar Pradesh, Arunachal Pradesh, Madhya Pradesh, Kerala, West Bengal, Jammu and Kashmir, Haryana, Mizoram, Punjab, Gujarat, Meghalaya, Sikkim and Bihar have already started large-scale demonstration activities and training programmes.

Investment Required

The Indian Government has made substantial investments in biotechnology research. Bringing Indian biotechnology products to market will require the involvement of large and small entrepreneurs and business houses. This will require substantial investments from Indian and overseas investors. The worldwide trend is that large companies are becoming major players in development of biotechnology products, and also in supporting product-related biotechnology re- search.

Expectations

In the years ahead, biotechnology R&D should produce a large number of new genetically improved plant varieties in India, including cotton, rice, brassicas, pigeonpea, mung bean, and wheat. Tissue culture regeneration protocols for important species such as mango, saffron, citrus, and neem will lead to major commercial activities. Micropropagation technology will provide high-quality planting materials to farmers. Environment-friendly biocontrol agents and biofertilizer packages will hopefully be made available to farmers in such a way that they can produce these in their own fields.

The country should be in a position to fully utilize, on a sustainable basis, medicinal and aromatic plants. The development through molecular biology of new diagnostic kits and vaccines for major diseases would make the health care system more efficient and cheaper. Genetic counselling clinics, molecular probes, and fingerprinting techniques should all be used to solve the genetic disorders in the population.

The establishment of ex situ gene banks to conserve valuable germplasm and diversity, and a large number of repositories, referral centers for animals, plants, and microorganisms should be possible. Detailed genetic readouts of individuals could be available. Information technology and biotechnology together should become a major economic force. It is expected that plants as bioreactors would be able to produce large numbers of proteins of therapeutic value, and many other important items. The recent discovery of the gene for recalcitrant species was a land- mark event. In vitro mass propagation can be carried out on any desired species with nonrandom programming. Certainly the 21st century could witness a major increase in new bioproducts generated through modern biology.

To achieve the goal of self-reliance in this field, India will require a strong educational and scientific base, clear public understanding of the value of new biotechnologies, and involvement of society in many of these biological ventures. India has a large research and educational infra- structure comprising 29 agriculture universities, 204 central and state universities, and more than 500 national laboratories and research institutions. It should therefore be possible to develop capabilities and programmes so that these institutions act as regional hubs for the farming community, where they can get direct feedback about new technological interventions. It will be equally important to establish strong partnerships and linkages with industry, from the time a research lead has emerged until the packaging of the technology and commercialization are achieved.

Arther Kornberg, Nobel Laureate, stated: "Much has been said about the future impact of biotechnology on industrial development, but this does not yet apply to the less developed countries that lack this infrastructure and industrial strength. In view of the current power of biotechnology and its even brighter future, there is no question that the less developed countries must now position and strengthen their status in biotechnology."

Kornberg further stressed that:"What a tragedy it would be if these enlarged concepts of genetics, biology and chemistry were available only to a small fraction of the world population located in a few major centres of highly developed countries."

Index

M

N

O

P

R

S

T

U

V